Nelson Mathematics 6

Nelson Mathematics 6

Series Authors and Senior Consultants
Mary Lou Kestell • Marian Small

Senior Authors
Heather Kelleher • Kathy Kubota-Zarivnij • Pat Milot
Betty Morris • Doug Super

Authors
Andrea Dickson • Jack Hope • Mary Lou Kestell
Pat Milot • Marian Small • Rosita Tseng Tam

Assessment Consultant
Damian Cooper

Australia   Canada   Mexico   Singapore   Spain   United Kingdom   United States

Nelson Mathematics 6

Series Authors and Senior Consultants
Mary Lou Kestell, Marian Small

Senior Authors
Heather Kelleher,
Kathy Kubota-Zarivnij, Pat Milot,
Betty Morris, Doug Super

Director of Publishing
Beverley Buxton

Publisher, Mathematics
Colin Garnham

Managing Editor, Development
David Spiegel

Senior Program Manager
Shirley Barrett

Program Managers
Colin Bisset
Mary Reeve

Developmental Editors
David Hamilton
Wendi Morrison
Bradley T. Smith
Bob Templeton
Susan Woollam

Developmental Consultants
Lynda Cowan
Jackie Williams

Editorial Assistants
Amanda Davis
Megan Robinson

Authors
Andrea Dickson, Jack Hope,
Mary Lou Kestell,
Pat Milot, Marian Small,
Rosita Tseng Tam

Executive Managing Editor, Special Projects
Cheryl Turner

Executive Managing Editor, Production
Nicola Balfour

Production Editor
Susan Selby

Copy Editor
Julia Cochrane

Indexer
Noeline Bridge

Senior Production Coordinator
Sharon Latta Paterson

Production Coordinators
Franca Mandarino
Kathrine Pummel

Creative Director
Angela Cluer

Art Director
Ken Phipps

Assessment Consultant
Damian Cooper

Art Management
ArtPlus Ltd., Suzanne Peden

Illustrators
ArtPlus Ltd., Deborah Crowle,
Sharon Matthews

Interior and Cover Design
Suzanne Peden

Cover Image
T. Kitchin/First Light

ArtPlus Ltd. Production Coordinator
Dana Lloyd

Composition
Heather Brunton/ArtPlus Ltd.

Photo Research and Permissions
Vicki Gould

Photo Shoot Coordinators
ArtPlus Ltd., Trent Photographics

Printer
Transcontinental Printing Inc.

COPYRIGHT © 2006 by Nelson, a division of Thomson Canada Limited.

Printed and bound in Canada
1 2 3 4 08 07 06 05

For more information contact Nelson, 1120 Birchmount Road, Toronto, Ontario, M1K 5G4. Or you can visit our Internet site at http://www.nelson.com

ALL RIGHTS RESERVED. No part of this work covered by the copyright hereon may be reproduced, transcribed, or used in any form or by any means—graphic, electronic, or mechanical, including photocopying, recording, taping, Web distribution, or information storage and retrieval systems—without the written permission of the publisher.

For permission to use material from this text or product, contact us by
Tel 1-800-730-2214
Fax 1-800-730-2215
www.thomsonrights.com

Every effort has been made to trace ownership of all copyrighted material and to secure permission from copyright holders. In the event of any question arising as to the use of any material, we will be pleased to make the necessary corrections in future printings.

Library and Archives Canada Cataloguing in Publication

Nelson mathematics 6 / Mary Lou Kestell, Marian Small.

Includes index.
ISBN 0-17-625971-6

1. Mathematics—Textbooks.
I. Kestell, Mary Louise II. Small, Marian III. Title: Nelson mathematics six.

QA107.2.N436 2005
510   C2005-903266-9

Advisory Panel

Senior Advisor

Doug Duff
Learning Supervisor
Thames Valley District School Board
London, Ontario

Advisors

Donna Anderson
Coal Tyee Elementary School
School District #68
Nanaimo-Ladysmith
Nanaimo, British Columbia

Keith Chong
Principal
School District #41
Burnaby, British Columbia

Attila Csiszar
Math Helping Teacher
Surrey School Board
Surrey, British Columbia

David P. Curto
Principal
Hamilton-Wentworth Catholic
District School Board
Hamilton, Ontario

Marg Curto
Principal of Programs, Elementary
Hamilton-Wentworth Catholic
District School Board
Hamilton, Ontario

Lillian Forsythe
Regina, Saskatchewan

Peggy Gerrard
Dr. Morris Gibson School
Foothills School Division
Okotoks, Alberta

Mary Gervais
Consultant
Durham Catholic District
School Board

C. Marie Hauk
Consultant
Edmonton, Alberta

Rebecca Kozol
School District #42
Maple Ridge, British Columbia

A. Craig Loewen
Associate Professor
University of Lethbridge
Lethbridge, Alberta

Frank A. Maggio
Department Head of Mathematics
Holy Trinity Catholic
Secondary School
Halton Catholic District
School Board
Oakville, Ontario

Moyra Martin
Principal
Calgary Catholic School District
Calgary, Alberta

Meagan Mutchmor
K–8 Mathematics Consultant
Winnipeg School Division
Winnipeg, Manitoba

Mary Anne Nissen
Consultant
Elk Island Public Schools
Sherwood Park, Alberta

Darlene Peckford
Principal
Horizon School Division #67
Taber, Alberta

Kathy Perry
Teacher
Peel District School Board
Brampton, Ontario

Susan Perry
Consultant
Durham Catholic District
School Board
Oshawa, Ontario

Bryan A. Quinn
Teacher
Edmonton Public Schools
Edmonton, Alberta

Ann Louise Revells
Principal
Ottawa-Carleton Catholic
School Board
Ottawa, Ontario

Lorraine Schroetter-LaPointe
Vice Principal
Durham District School Board
Oshawa, Ontario

Nathalie Sinclair
Assistant Professor
Michigan State University
East Lansing, Michigan

Susan Stuart
Assistant Professor
Nipissing University
North Bay, Ontario

Doug Super
Principal
Vancouver School Board
Vancouver, British Columbia

Joyce Tonner
Learning Supervisor
Thames Valley District
School Board
London, Ontario

Stella Tossell
Mathematics Consultant
North Vancouver, British Columbia

Gerry Varty
AISI Math Coordinator
Wolf Creek School Division #72
Ponoka, Alberta

Michèle Wills
Assistant Principal
Calgary Board of Education
Calgary, Alberta

Reviewers

Michael Babcock
Limestone District School Board

Nancy Campbell
Rainbow District School Board

Catherine Chau
Toronto District School Board

Deb Colvin-MacDormand
Edmonton Public Schools

William Corrigan
Lakeshore School Board

Anna Dutfield
Toronto District School Board

Susan Gregson
Peel District School Board

Wendy King
Coley's Point Primary School

Gowa Kong
North Vancouver School District

Joan McDuff
Faculty of Education
Queen's University

Ken Mendes
Ottawa-Carleton Catholic
School Board

Sandie Rowell
Hamilton-Wentworth District
School Board

Rose Scaini
York Catholic District School Board

Lindy Smith
Peel District School Board

Triona White
Ottawa-Carleton Catholic
School Board

Aboriginal Content Reviewers

Brenda Davis
Education Consultant
Six Nations

Laura Smith
Educational Consultant
Abbotsford, British Columbia

Equity Reviewer

Mary Schoones
Educational Consultant/
Retired Teacher
Ottawa-Carleton District
School Board

Literacy Reviewer

Roslyn Doctorow
Educational Consultant

Thank you to the following teachers for testing the Chapter Tasks.

Jason Chenier
Peel District School Board

Kate Pratt
Peel District School Board

Marjolijne Vanderkrift
Peel District School Board

Contents

CHAPTER 1

Patterns in Mathematics — 1

- **Getting Started:** Growth Patterns — 2
- **Lesson 1:** Writing Pattern Rules — 4
- **Curious Math:** Math Magic — 7
- **Lesson 2:** Relationships Rules for Patterns — 8
- **Mental Math:** Pairing to Multiply — 11
- **Lesson 3:** Variables in Expressions — 12
- **Lesson 4:** Representing Patterns on a Graph — 14
- **Math Game:** Who Am I? — 17
- **Frequently Asked Questions** — 18
- **Mid-Chapter Review** — 19
- **Lesson 5:** Patterns and Spreadsheets — 20
- **Curious Math:** Rice on a Chessboard — 21
- **Lesson 6:** Solve a Simpler Problem — 22
- **Lesson 7:** Equal Expressions — 24
- **Lesson 8:** Variables in Equations — 26
- **Skills Bank** — 28
- **Problem Bank** — 30
- **Frequently Asked Questions** — 32
- **Chapter Review** — 33
- **Chapter Task:** Patterns in Your Life — 34

Rodrigo

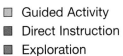

- Guided Activity
- Direct Instruction
- Exploration

CHAPTER 2

Numeration — 35

Getting Started: Number Clues	36
Lesson 1: Exploring Greater Numbers	38
Curious Math: Billions	39
Lesson 2: Reading and Writing Numbers	40
Lesson 3: Comparing and Ordering Numbers	42
Lesson 4: Renaming Numbers	44
Lesson 5: Communicate about Solving Problems	46
Frequently Asked Questions	48
Mid-Chapter Review	49
Lesson 6: Reading and Writing Decimal Thousandths	50
Math Game: Close as You Can	53
Lesson 7: Rounding Decimals	54
Lesson 8: Comparing and Ordering Decimals	56
Mental Math: Dividing Decimals by Renaming	58
Skills Bank	59
Problem Bank	62
Frequently Asked Questions	63
Chapter Review	64
Chapter Task: Reporting Numbers	66

Li Ming

CHAPTER 3

Data Management — 67

- **Getting Started:** Memorizing Pictures — 68
- **Lesson 1:** Creating and Analyzing a Survey — 70
- **Mental Math:** Determining Missing Decimals — 71
- **Lesson 2:** Plotting Coordinate Pairs — 72
- **Lesson 3:** Line Graphs — 74
- **Curious Math:** Telling Stories about Graphs — 77
- **Lesson 4:** Scatter Plots — 78
- **Math Game:** 4 in a Row — 81
- **Frequently Asked Questions** — 82
- **Mid-Chapter Review** — 83
- **Lesson 5:** Mean and Median — 84
- **Lesson 6:** Changing the Intervals on a Graph — 86
- **Lesson 7:** Changing the Scale on a Graph — 87
- **Lesson 8:** Communicate about Conclusions from Data Displays — 88
- **Lesson 9:** Constructing Graphic Organizers — 90
- **Skills Bank** — 93
- **Problem Bank** — 95
- **Frequently Asked Questions** — 97
- **Chapter Review** — 98
- **Chapter Task:** Investigating Body Relationships — 100
- **CHAPTERS 1–3 CUMULATIVE REVIEW** — 101

Denise

NEL

CHAPTER 4

Addition and Subtraction 105

Getting Started: Planning Cross-Country Routes — 106

Lesson 1: Adding and Subtracting Whole Numbers — 108

Curious Math: Number Reversal — 109

Lesson 2: Estimating Sums and Differences — 110

Lesson 3: Adding Whole Numbers — 112

Lesson 4: Subtracting Whole Numbers — 114

Curious Math: Subtracting a Different Way — 117

Frequently Asked Questions — 118

Mid-Chapter Review 104 — 119

Lesson 5: Adding and Subtracting Decimals — 120

Math Game: Mental Math with Money — 121

Lesson 6: Adding Decimals — 122

Lesson 7: Subtracting Decimals — 124

Mental Math: Using Whole Numbers to Add and Subtract Decimals — **127**

Lesson 8: Communicate about Solving a Multi-Step Problem — 128

Skills Bank — 130

Problem Bank — 133

Frequently Asked Questions — 135

Chapter Review — 136

Chapter Task: Gold Coins — 138

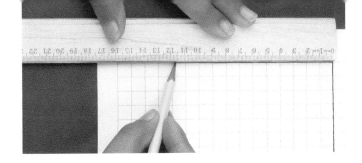

CHAPTER 5

Measuring Length — 139

Getting Started: Racing Snails	140
Lesson 1: Measuring Length	142
Lesson 2: Metric Relationships	144
Lesson 3: Perimeters of Polygons	146
Frequently Asked Questions	148
Mid-Chapter Review	149
Lesson 4: Solve Problems Using Logical Reasoning	150
Curious Math: Triangle Sides	152
Math Game: Lines, Lines, Lines	153
Lesson 5: Exploring Perimeter	154
Mental Math: Calculating Lengths of Time	155
Skills Bank	156
Problem Bank	158
Frequently Asked Questions	159
Chapter Review	160
Chapter Task: Mapping a Triathlon Course	162

Jorge

CHAPTER 6

Multiplication and Division — 163

Getting Started: Recycling Milk Cartons — **164**

Lesson 1: Identifying Factors, Primes, and Composites — 166

Curious Math: Separating Primes from Composites — **169**

Lesson 2: Identifying Multiples — 170

Lesson 3: Calculating Coin Values — 172

Mental Math: Halving and Doubling to Multiply — **173**

Lesson 4: Multiplying by Hundreds — 174

Lesson 5: Estimating Products — 176

Lesson 6: Multiplying by Two-Digit Numbers — 178

Curious Math: Lattice Multiplication — **181**

Frequently Asked Questions — **182**

Mid-Chapter Review — **183**

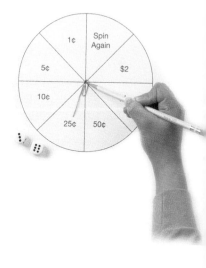

Lesson 7: Dividing by 1000 and 10 000 — 184

Lesson 8: Dividing by Tens and Hundreds — **186**

Lesson 9: Estimating Quotients — 188

Lesson 10: Dividing by Two-Digit Numbers — **190**

Math Game: Coin Products — **193**

Lesson 11: Communicate About Creating and Solving Problems — 195

Lesson 12: Order of Operations — 196

Curious Math: Egyptian Division — **197**

Skills Bank — **198**

Problem Bank — **201**

Frequently Asked Questions — **203**

Chapter Review — **204**

Chapter Task: Chartering Buses — **206**

CHAPTER 7

2-D Geometry — 207

Getting Started: Mystery Design	208
Lesson 1: Estimating Angle Measures	210
Lesson 2: Investigating Properties of Triangles	212
Mental Imagery: Visualizing Symmetrical Shapes	213
Lesson 3: Communicate About Triangles	214
Frequently Asked Questions	216
Mid-Chapter Review	217
Lesson 4: Constructing Polygons	218
Curious Math: Folding Along Diagonals	221
Lesson 5: Sorting Polygons	222
Lesson 6: Investigating Properties of Quadrilaterals	224
Skills Bank	226
Problem Bank	228
Frequently Asked Questions	229
Chapter Review	230
Chapter Task: Furnishing a Bedroom	232
CHAPTERS 4–7 CUMULATIVE REVIEW	233

Tara

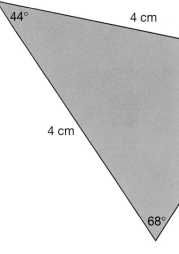

CHAPTER 8

Area — 237

Getting Started: Area Puzzle	**238**
Lesson 1: Unit Relationships	240
Lesson 2: Area Rule for Parallelograms	242
Mental Imagery: Estimating Area	**244**
Lesson 3: Geometric Relationships	245
Lesson 4: Area Rule for Triangles	246
Frequently Asked Questions	**248**
Mid-Chapter Review	**249**
Lesson 5: Solve Problems Using Equations	250
Lesson 6: Areas of Polygons	252
Curious Math: Changing Parallelograms	**254**
Skills Bank	**255**
Problem Bank	**257**
Frequently Asked Questions	**259**
Chapter Review	**260**
Chapter Task: Design a Placemat and Napkin Set	**262**

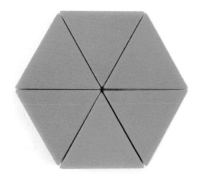

Raven

Akeem

Chandra

CHAPTER 9

Multiplying Decimals — 263

Getting Started: Downloading Songs	264
Lesson 1: Estimating Products	266
Mental Math: Multiplying by 5 and 50	267
Lesson 2: Multiplying by 1000 and 10 000	268
Lesson 3: Mulitplying Tenths by Whole Numbers	270
Frequently Asked Questions	272
Mid-Chapter Review	273
Lesson 4: Multiplying by 0.1, 0.01, or 0.001	274
Lesson 5: Multiplying Multiples of Ten by Tenths	276
Lesson 6: Communicate about Problem Solving	278
Lesson 7: Choosing a Multiplication Method	280
Curious Math: Decimal Equivalents	281
Math Game: Race to 50	282
Skills Bank	283
Problem Bank	286
Frequently Asked Questions	287
Chapter Review	288
Chapter Task: Growing Up	290

CHAPTER 10

Dividing Decimals 291

Getting Started: Ancient Length Measurements — 292

Lesson 1: Estimating Quotients — 294

Math Game: Estimate the Range — 297

Lesson 2: Dividing Money — 298

Mental Math: Adding Decimals by Renaming — 299

Lesson 3: Dividing Decimals by One-Digit Numbers — 300

Curious Math: Dividing Magic Squares — 303

Frequently Asked Questions — 304

Mid-Chapter Review — 305

Lesson 4: Dividing by 10, 100, 1000, and 10 000 — 306

Math Game: Calculate the Least Number — 309

Lesson 5: Solve Problems by Working Backward — 310

Skills Bank — 312

Problem Bank — 315

Frequently Asked Questions — 316

Chapter Review — 317

Chapter Task: Judging Fairness — 318

Tom

CHAPTER 11

3-D Geometry and 3-D Measurement — 319

Getting Started: Solving Net Puzzles	320
Lesson 1: Visualizing and Constructing Polyhedrons	322
Mental Imagery: Drawing Faces of Polyhedrons	323
Lesson 2: Surface Area of Polyhedrons	324
Lesson 3: Volume of Rectangular and Triangular Prisms	326
Curious Math: Cross-Sections	329
Lesson 4: Solve Problems by Making a Model	330
Frequently Asked Questions	332
Mid-Chapter Review	333
Lesson 5: Creating Isometric Sketches	334
Lesson 6: Creating Cube Structures from Sketches	336
Lesson 7: Different Views of a Cube Structure	338
Lesson 8: Creating Cube Structures from Different Views	340
Curious Math: Plane of Symmetry	341
Skills Bank	342
Problem Bank	344
Frequently Asked Questions	345
Chapter Review	346
Chapter Task: Painting Bids	348
CHAPTERS 8–11 CUMULATIVE REVIEW	349

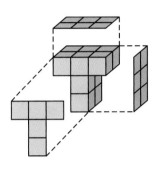

CHAPTER 12

Fractions, Decimals, and Ratios — 353

Getting Started: Filling a Pancake Order	354
Lesson 1: Comparing and Ordering Fractions	356
Lesson 2: Comparing Fractions with Unlike Denominators	358
Lesson 3: Fraction and Decimal Equivalents	360
Mental Math: Using Factors to Multiply	361
Lesson 4: Ratios	362
Lesson 5: Equivalent Ratios	364
Frequently Asked Questions	366
Mid-Chapter Review	367
Lesson 6: Percents as Special Ratios	368
Lesson 7: Relating Percents to Decimals and Fractions	370
Lesson 8: Estimating and Calculating Percents	372
Lesson 9: Unit Rates	374
Lesson 10: Solving Problems Using Guess and Test	376
Math Game: Ratio Concentration	378
Skills Bank	379
Problem Bank	382
Frequently Asked Questions	383
Chapter Review	384
Chapter Task: Running with Terry	386

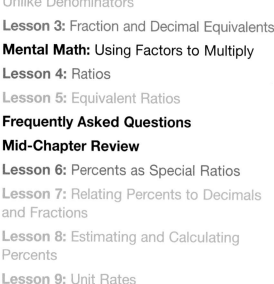

James

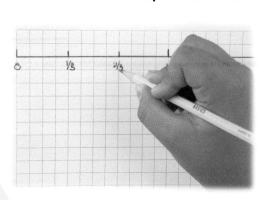

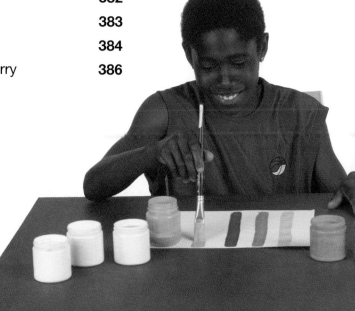

CHAPTER 13

Probability — 387

- **Getting Started:** Lucky Seven — 388
- **Lesson 1:** Conducting Probability Experiments — 390
- **Mental Imagery:** Visualizing Fractions on a Number Line — 391
- **Lesson 2:** Using Percents to Describe Probabilities — 392
- **Lesson 3:** Solving a Problem by Conducting an Experiment — 394
- **Curious Math:** Random Numbers and Letters — 397
- **Frequently Asked Questions** — 398
- **Mid-Chapter Review** — 399
- **Lesson 4:** Theoretical Probability — 400
- **Lesson 5:** Tree Diagrams — 402
- **Lesson 6:** Comparing Theoretical and Experimental Probability — 404
- **Math Game:** No Tails Please! — 406
- **Skills Bank** — 407
- **Problem Bank** — 409
- **Frequently Asked Questions** — 411
- **Chapter Review** — 412
- **Chapter Task:** Winning Races — 414

Kurt

CHAPTER 14

Patterns and Motion in Geometry — 415

Getting Started: Creating a Design Using Transformations	**416**
Lesson 1: Describing Rotations	418
Lesson 2: Performing and Measuring Rotations	420
Lesson 3: Rotational Symmetry	**422**
Curious Math: Alphabet Symmetry	**425**
Frequently Asked Questions	**426**
Mid-Chapter Review	**427**
Lesson 4: Communicate Using Diagrams	428
Lesson 5: Exploring Transformation Patterns with Technology	430
Lesson 6: Creating Designs	432
Mental Imagery: Identifying Transformations	**434**
Skills Bank	**435**
Problem Bank	**437**
Frequently Asked Questions	**438**
Chapter Review	**439**
Chapter Task: Creating Nets	**440**
CHAPTERS 12–14 CUMULATIVE REVIEW	**441**
GLOSSARY	**445**
ANSWERS TO SELECTED PROBLEMS	**465**
INDEX	**466**
CREDITS	**473**

CHAPTER 1

Patterns in Mathematics

Goals

You will be able to

- use models and tables to represent patterns
- identify, extend, and create patterns
- analyze, represent, and describe patterns
- use patterns to solve problems

Everybody shakes hands

CHAPTER 1

Getting Started

Growth Patterns

Fardad got a growth chart as a gift on his 5th birthday. He kept it on his closet door for five years and recorded his height every year on his birthday.

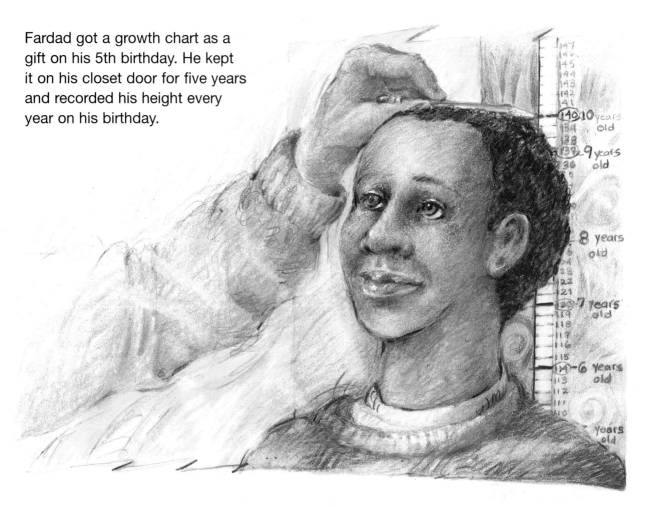

? **What patterns are there in the average heights of males and females?**

A. Describe the patterns you see in Fardad's growth chart.

B. Write the pattern rule for Fardad's height from ages 5 to 8.

C. If you were 109 cm at age 5 and you grew by 5 cm a year, how tall would you be now? Represent your pattern in a table of values and write a pattern rule.

Age (in years)	Fardad's height (cm)
5	108
6	114
7	120
8	126
9	137
10	140

D. Describe the patterns you see for the average height of boys in this chart.

E. Compare the patterns in Fardad's chart to the patterns you saw in part D.

F. Describe the patterns you see for the average height of girls in this chart.

G. Compare the growth patterns of boys and girls.

Age (in years)	Average height of boys (cm)	Average height of girls (cm)
5	110	109
6	116	115
7	122	121
8	128	127
9	134	134
10	140	140
11	148	146
12	154	152
13	160	158
14	166	161
15	172	164

Do You Remember?

1. Write the next three numbers in each pattern.
   a) 15, 30, 45, 60, …
   b) 144, 133, 122, 111, …
   c) 1, 11, 111, 1111, …
   d) 2, 5, 9, 14, …
   e) 97, 88, 78, 67…

2. Describe each pattern in Question 1. Write pattern rules for Parts a) and b).

3. a) Make the next design in this pattern.
   b) Record the number of green triangles and the number of red trapezoids in each design. Use a table.
   c) How many of each shape are in design 6?

design 1   design 2   design 3

4. Grace planned a welcome back party for her friends. She made this table to decide how much food to order.
   a) How many bags of chips would she order for 18 people?
   b) Grace ordered 60 items. How many people were at the party?

Number of people	Number of pizzas	Number of bags of chips	Number of bottles of juice
3	1	2	3
6	2	4	6
9	3	6	9

CHAPTER 1

1 Writing Pattern Rules

You will need
- a calculator

Goal Use rules to extend patterns and write pattern rules.

Planet	Length of 1 day (in hours)
Earth	24
Neptune	16

In 100 h, the conditions should be good for Denise to see Neptune with her telescope. She wonders how many days that is on Earth and on Neptune.

? How many days is 100 h on Earth and on Neptune?

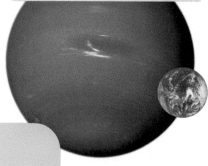

Denise's Solution

I started to make a table, but I can use patterns to answer the question.

The pattern in the Number of hours on Earth column is 24, 48, 72, ...

Number of days	Number of hours on Earth	Number of hours on Neptune
1	24	16
2	48	32
3	72	48

The 1st term in the pattern is $1 \times 24 = 24$.

The 2nd term in the pattern is $2 \times 24 = 48$.

To calculate the number of hours on Earth in 4 days, I can multiply 24 by the **term number** 4.

$4 \times 24 = 96$

The 4th term in the number of hours on Earth pattern is 96. 4 days are 96 hours on Earth.

I can write a pattern rule using the term and term number:

Pattern rule: Each term is 24 multiplied by the term number.

term number
A number that tells the position of a term in a pattern.
1, 3, 5, 7, ...

1st term   2nd term

5 is the 3rd term or term number 3

A. Write a rule like Denise's for the number of hours on Neptune pattern.

B. Calculate the fourth term in the number of hours on Neptune pattern.

C. Estimate the number of terms you will need to be past 100 h for Earth.

D. Estimate the number of terms you will need to be past 100 h for Neptune.

E. About how many days is 100 h on Earth and on Neptune?

Reflecting

1. a) How can you use Denise's rule to determine the number of hours that would have passed on Earth in 10 days?
   b) How would you use your rule to determine the number of hours that would have passed on Neptune in 20 days?

2. In the past, you would have written a pattern rule by describing how the pattern starts and how it can be extended.
   a) For the Earth hours pattern, describe how the starting number and the number that describes how you extend the pattern are related.
   b) Is this true for the Neptune hours pattern? Explain.

Checking

3. It takes almost 10 h for 1 day on Jupiter.
   a) Write the first few terms in the pattern for the number of hours in Jupiter days.
   b) Write a pattern rule like Denise's for Jupiter. Use the rule to determine the 15th term in the Jupiter pattern.

Practising

4. a) Complete this table for Mars.
    b) Write a pattern rule that will tell you the number of hours on Mars for any number of days on Mars.
    c) Use your pattern rule to determine the approximate number of hours in seven days on Mars.

Mars

Number of days	Number of hours
1	24.75
2	
3	

5. One astronomical unit (1 AU) is the average distance from the Sun to Earth. 1 AU is about 150 million kilometres. It takes light 0.14 h to travel 1 AU.
    a) Write a pattern rule for calculating a term in the pattern for the time light takes to travel any number of astronomical units.
    b) It is about 5 AU from the Sun to Jupiter. About how many hours does it take light to travel from the Sun to Jupiter?
    c) It is 30 AU from the Sun to Neptune. About how many hours does it take light to travel from the Sun to Neptune?

Number of astronomical units	Time light takes to travel (in hours)
1	0.14
2	0.28
3	0.42
4	0.56

6. Samantha is going to space camp next summer. She is saving $14 each month to pay for it.
    a) Write a pattern of numbers that shows how much she saves in one to five months.
    b) Write a pattern rule for determining how much she saves in any number of months.
    c) Determine the amount of money Samantha saves in eight months.
    d) The space camp costs $125. For how many months does Samantha need to save?

Curious Math

Math Magic

Students often enjoy impressing their friends with magic tricks. Here are two magic tricks that use operations with numbers.

Magic Trick 1: Lucky 13

Steps	Directions
1	Choose a two-digit number. Record it secretly.
2	Add 10.
3	Double the number.
4	Add 6.
5	Divide by 2.
6	Subtract the original number.
7	Record your answer.

1. Choose a number and follow the steps in Lucky 13. Choose some more numbers. What do you notice?

Magic Trick 2: You Get What You Give

Steps	Directions
1	Choose a three-digit number. Record it secretly.
2	Add 10.
3	Triple your number.
4	Subtract the original number.
5	Subtract 23.
6	Subtract the number of days in a week.
7	Divide by 2.
8	Record your answer.

2. Choose a number and follow the steps in You Get What You Give. Choose some more numbers. What do you notice?

CHAPTER 1

2 Relationship Rules for Patterns

You will need
- rhombus pattern blocks
- a calculator

Goal Write relationship pattern rules based on the term number.

Qi's family business makes square-dancing outfits. Each shirt and skirt is made up of sections, and each section has a zig-zag design that is outlined with cord.

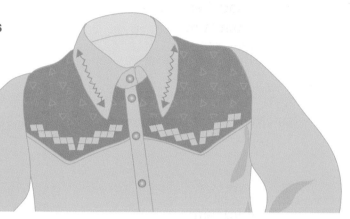

? What length of cord is needed for a section with 20 rhombuses in its design?

Qi's Solution

I model the zig-zag with pattern blocks and make a table to keep track of the perimeter in each figure. One side of one rhombus is 1 unit or 1 cm.

    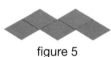

figure 1   figure 2   figure 3   figure 4   figure 5

I complete a table of values.

Term number (figure number)	Perimeter
1	4
2	6
3	8
4	10

The perimeter pattern is 4, 6, 8, 10, …
I need to determine the 20th term.

I can write a **recursive pattern rule** to describe how the pattern starts and how to extend it. "Start with 4 and add 2 each time." But to use this rule to determine the 20th term, first I'd have to determine the 19th term, and the 18th term before that, … right back to the start.

I'd like to determine the 20th term without all that calculation, so I'm going to use an **explicit pattern rule**.

There is a difference of 2 between any two terms in this pattern.

The **common difference** is 2.

The 2nd term is 4 + 2 = 6.

The 3rd term is 4 + two 2s = 4 + 4 = 8.

The 4th term is 4 + three 2s = 4 + 6 = 10.

Explicit pattern rule: Start with the 1st term and add 2 one time less than the term number.

I will calculate the 20th term.

20th term = 4 + nineteen 2s
         = 4 + 38
         = 42

I will need 42 cm of cord.

recursive pattern rule
A pattern rule that tells you the start number of a pattern and how the pattern continues.

"Start with 5 and add 3" is a recursive pattern rule for 5, 8, 11, 14, …

explicit pattern rule
A pattern rule that uses the term number to determine a term in the pattern

common difference
The difference between any two consecutive terms in a pattern.

3, 7, 11, 15, …
15 − 11 = 4 and
11 − 7 = 4
This pattern has a common difference of 4.

Reflecting

1. How could you use Qi's explicit pattern rule to determine the 15th term of the perimeter pattern?

2. If a pattern grows by the same amount every term, why is it fairly easy to describe it using an explicit pattern rule?

3. What is the advantage of using a recursive pattern rule? What is the advantage of using an explicit pattern rule?

Checking

4. At a restaurant, chairs and tables are put together as shown.
   a) Complete the table up to the 7th arrangement.

Term number (arrangement number)	Number of chairs
1	5
2	8

   1st arrangement

   b) What is the common difference in the number of chairs pattern?
   c) What is the eighth term for the number of chairs pattern? Explain.
   d) Determine the number of chairs in the 20th arrangement. Use a pattern rule. Show your work.

   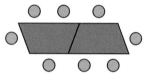
   2nd arrangement

   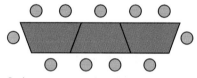
   3rd arrangement

Practising

5. a) Complete a table to show the number of geese in the 1st to the 4th arrangements.
   b) Write the 1st term and the common difference.
   c) How many geese are in the 10th arrangement? Use a pattern rule. Show your work.

6. Determine the 10th term in each pattern. Use a pattern rule. Show your work.
   a) 3, 6, 9, 12, …
   b) 29, 38, 47, 56, …
   c) 10, 15, 20, 25, …
   d) 5, 10, 15, 20, …
   e) 1.5, 3.0, 4.5, 6.0, …
   f) $5.00, $5.50, $6.00, $6.50, …

   1st arrangement

7. A ticket to the fair costs $2.00 on September 1. Each day, the price increases by $0.25.
   a) Write the first four terms in the price pattern.
   b) What is the common difference of the pattern?
   c) What is the price of a ticket on September 12? Use a pattern rule. Show your work.

   2nd arrangement

8. Kevin is saving $2.00 a week each week from the start of school until December break.
   a) Write the first four terms of his savings totals in a pattern.
   b) What is the common difference of the pattern?
   c) How much will he have saved by the 12th week? the 16th week? Use a pattern rule. Show your work.

   3rd arrangement

9. The 1st arrangement of a marching band has seven players. Four players join and they have 11 players in the second arrangement. They add four players to each new arrangement. What arrangement has 55 players marching?

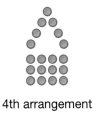

4th arrangement

Mental Math

Pairing to Multiply

Sometimes you can calculate the product of several numbers by pairing numbers that are easy to multiply.

To calculate $2 \times 13 \times 50$, you can pair 2 and 50 because then you can use mental math to calculate the entire product.

$2 \times 13 \times 50$
100
$100 \times 13 = 1300$

A. Why is it easier to multiply 2×50 and then $\times 13$ rather than 2×13 and then $\times 50$?

Try These

1. a) $2 \times 9 \times 5$
   b) $7 \times 5 \times 20$
   c) $4 \times 8 \times 50$
   d) $2 \times 6 \times 150$
   e) $20 \times 20 \times 5 \times 5$
   f) $8 \times 9 \times 5$

2. a) $2 \times 157 \times 0.5$
   b) $7 \times 5 \times 0.4$
   c) $4 \times 15.6 \times 2.5$
   d) $0.2 \times 18.2 \times 5$
   e) $25 \times 0.5 \times 8$
   f) $0.8 \times 9 \times 5$

3. Determine the missing numbers.
   a) $2 \times \blacksquare \times 5 = 120$
   b) $\blacksquare \times 7 \times 5 = 70$
   c) $5 \times 17 \times \blacksquare = 170$
   d) $2 \times \blacksquare \times 5 = 320$
   e) $2 \times 2 \times 9 \times \blacksquare \times \blacksquare = 900$
   f) $\blacksquare \times \blacksquare \times 7 \times 5 \times 5 = 700$

CHAPTER 1

3 Variables in Expressions

You will need
- a calculator

Goal Use variables in an expression.

Isabella and Jorge are placing markers along a cross-country running route. They need to place the 1st marker 50 m from the start and place the others every 50 m to the end of the 2 km route.

? How many markers do they need to place?

Isabella's Solution

The distance in metres to the 1st marker is 1×50.

The distance in metres to the 4th marker is 4×50.

The number of markers we place is a **variable**. I will use n to represent the variable. Now I can write an **expression** to show the distance to any marker.

variable
A quantity that varies or changes.
A variable is often represented by a letter or a symbol.

expression
A mathematical statement made with numbers or variables and symbols—
$5 + 3 - 7$ is an expression

NEL

The distance in metres to the nth marker is $n \times 50$.

To calculate how far out each marker is, I can replace n with the number of the marker.

The distance in metres to the 8th marker is $8 \times 50 = 400$.

The 8th marker is 400 m from the start.

Communication Tip

When you represent a variable with a letter, you might want to choose the first letter of the variable. This will help you to remember what the letter stands for. For example, Isabella chose n because it stood for "Number of markers." She could have chosen m for "markers".

A. Calculate the distances at which Isabella and Jorge should place the first four markers.

B. Replace n in Isabella's expression with the values 1, 2, 3, and 4. Do you agree that Isabella's expression tells the distance of each marker from the start?

C. Determine the distance, in metres, from the start of the route to the location of the last marker.

D. Try different values of n in Isabella's expression. Show your work. Which value of n gives the distance to the last marker?

E. How many markers do they have to place?

Reflecting

1. Why did Isabella say that the number of markers they place is a variable?

2. Isabella said, "If we placed markers 20 m apart, the distance to the nth marker would be $20 \times n$." Could you replace n in Isabella's sentence with any number? For example, could you replace it with $\frac{1}{2}$? Explain.

3. To calculate the distance to the 15th marker you could add 50 + fourteen more 50s". How could you use a variable to write a pattern rule to calculate any value in the pattern of distances to markers?

4 Representing Patterns on a Graph

You will need
- grid paper
- a calculator

Goal Represent patterns in tables and on graphs.

Khaled is playing a video game called *Real Life*. In his current "job," he earns two tokens for each task that he does. In the game, one piece of clothing costs 16 tokens.

? How many tasks does Khaled need to do to earn 16 tokens?

Khaled's Solution

I used the pattern rule $n \times 2$ to determine the number of tokens I need to earn. I recorded values in a table and graphed the data.

Number of tasks, n	Pattern rule $n \times 2$	Total tokens earned
0	0×2	0
1	1×2	2
2	2×2	4
3	3×2	6
4	4×2	8

I put the number of tasks on the horizontal axis and the total tokens earned on the vertical axis.

I graphed five points: (0, 0), (1, 2), (2, 4), (3, 6), and (4, 8).

For example, to plot the point for 4 tasks and 8 tokens, I started at 0 on the axes and moved 4 to the right and 8 up. I noticed the points formed a line, so I joined them with a dotted line.

Total Tokens Earned Compared to Number of Tasks

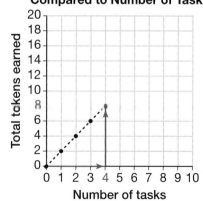

I extended the dotted line. Then I went to 16 on the Total tokens earned axis and estimated where the point at that height would be on the line. I looked down and read the value on the Number of tasks axis. This point is at eight tasks, so I need to do eight tasks to earn 16 tokens.

I checked this number with my pattern rule. The number of tokens earned for 8 tasks is $n \times 2 = 8 \times 2 = 16$.

The answer checks. I have to do 8 tasks to earn 16 tokens.

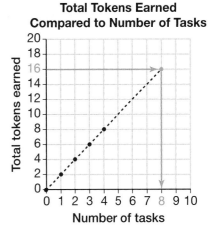

Reflecting

1. In Khaled's pattern rule, what does the variable n represent?

2. How did the graph help Khaled to know how many tasks he had to do to earn 16 tokens?

3. Why is solving Khaled's problem using a graph actually solving the number of tasks required to earn any number of tokens, not just 16?

Checking

4. Li Ming is playing *Real Life*. She has saved 20 000 tokens and wants to buy furniture for her house. Each piece of furniture costs 2000 tokens. This graph shows how her savings would decrease if she bought 1, 2, or 3 pieces of furniture.

   a) Use the graph to determine how many tokens Li Ming would have if she bought 5 pieces of furniture.
   b) Describe the pattern for the number of tokens Li Ming has left.
   c) How many pieces of furniture can she buy and still have 4000 tokens?

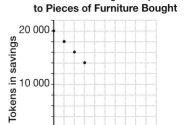

Practising

5. Nathan is buying juice boxes in packages of three.
   a) Make a table to show the total number of Nathan's juice boxes for 1 to 10 packages.
   b) Graph the total number of juice boxes compared to the total number of packages.
   c) Describe the graph.
   d) How many packages does Nathan need to buy so 24 people get one box each?

6. Simone is buying juice boxes in packages of four.
   a) Make a table to show the total number of Simone's juice boxes for 1 to 10 packages.
   b) Graph the total number of juice boxes compared to the total number of packages. Use the same graph you used in Question 6, but use a different colour.
   c) Describe the graph.
   d) How many packages does Simone need to buy so 24 people can have one box each?

7. Terry earns $8 for each hour she works at the corner store.
   a) Copy and complete the table.
   b) Graph Terry's total earnings compared to the number of hours Terry works.
   c) Determine the number of hours Terry has to work to earn $75.

Number of hours, h	Pattern rule $h \times 8$	Total earnings
0		
1		
2		
3		
4		

8. In the game Real Life, Ivan has saved 20 000 tokens. He needs to pay 4500 tokens per month for food and rent. He wants to quit working and learn a new skill. He will pay for food and rent with his savings.
   a) Make a table to show the total number of tokens Ivan has for one to six months.
   b) Graph the total number of tokens compared to the number of months. Plot at least four points on the graph. Connect the points with a dotted line.
   c) For how many months will Ivan's savings last?

Math Game

Who Am I?

Number of players: 2

> **You will need**
> - Who Am I? cards

Play with 8 cards

How to play: Identify the card your partner has picked as the challenge card.

Step 1   Your partner picks one card to be the challenge card and secretly records the attributes of the person on that card.

Step 2   You arrange the cards face up on the table and then ask questions with yes-or-no answers. Turn the cards that you eliminate face down. When you think you know which card your partner chose, point to it. If you are wrong, guess again. If you are right, your turn is over. Record the number of guesses you used to identify the challenge card.

 ### Emilio's Turn

The person on the card Tara picked has glasses and a watch, but no hat.

I ask, "Do you have a hat?"

Tara says, "No."

I turn down all the cards that have hats.

I ask, "Do you have glasses?"

Tara says, "Yes."

I turn down all the cards that do not have glasses.

I look at my remaining cards.

I guess that my opponent's hidden card has glasses, no hat, and a watch. I point to that card.

Tara looks at the hidden card and says, "You're right."

I record that it took me 1 guess.

no hat
glasses
watch

CHAPTER 1

Frequently Asked Questions

Q: How can you use a pattern rule to calculate term values for a pattern that increases or decreases by a constant amount?

A1: You can use a recursive rule to continue the pattern to that term by knowing what to add or subtract from the previous term. For example, to calculate what is the 8th term of 6, 9, 12, 15, ..., continue the pattern by adding 3 to each term: 6, 9, 12, 15, 18, 21, 24, 27. The 8th term is 27.

A2: You can use an explicit pattern rule: add the 1st term to the product of the common difference and 1 less than the term number you want. For example, what is the 8th term of 6, 9, 12, 15, ...? The 1st term is 6 and the common difference is 3.

8th term $= 6 +$ seven 3s
$= 6 + 21$
$= 27$

Q: How do you calculate the value of an expression that includes a variable?

A: Replace the symbol with a value of the variable, then calculate as usual. For example, each soccer player needs two shoelaces. How many shoelaces do 13 players need?

Number of laces $= p \times 2$ — This expression tells how many shoelaces p players need. "p" represents the variable "number of players."

The expression says, "The number of laces equals two times the number of players".

$= 13 \times 2$ — To calculate the number of laces for 13 players, replace p with 13. Calculate normally. 13 players need 26 laces.

$= 26$ — 13 players need 26 laces.

Q. How do you graph a pattern?

A. Make a table of values. Draw a horizontal axis for the term numbers and a vertical axis for the term values. To graph a point, start where the axes meet, at 0. Count to the right to the term number and then up to the term value. Draw a dot. Join the dots. For example, here is a graph for 4, 7, 10, 13, 16, ...

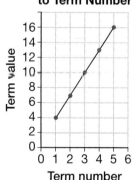

Term Value Compared to Term Number

CHAPTER 1
Mid-Chapter Review

LESSON

1
1. As a space station orbits Earth, astronauts see 15 sunrises in 24 h.
   a) Copy and complete this table for up to five Earth days.

Number of Earth days	Number of sunrises seen on the space station
1	15
2	
3	

   b) How many sunrises would an astronaut see in seven Earth days?
   c) Shannon Lucid spent 188 Earth days on the space station. How many sunrises could she have seen? Use a pattern rule. Show your work.

2
2. Tristen has $200 in his savings account. Starting next week, he plans to add $10 each week.
   a) Make a table to show Tristen's balance after five weeks.
   b) Write a pattern rule that tells how to calculate his balance after any number of weeks.
   c) How much will be in Tristen's account after 52 weeks?

3
3. This sentence describes how many days an astronaut is in space. Earth days = $h \div 24$. The symbol h represents the variable 'number of hours in space.' Calculate the number of Earth days an astronaut is in space for each value of h.
   a) 48   c) 100
   b) 96   d) 600

4
4. Kristen has a notebook with 64 pages. Each day she uses six pages of the book. On what day will she have less than half of the notebook left? Use a graph.

CHAPTER 1

5 Patterns and Spreadsheets

You will need
- spreadsheet software

Goal: Create patterns using spreadsheets and compare the growth.

Chandra saved the local rajah's elephants, so the mean rajah offered her a reward. He gave her a choice of beautiful jewels, but she noticed a chessboard. "All I ask for is rice," she said with a smile. "Please put two grains of rice on the first square of the chessboard. Put four grains on the second square, put eight on the next, and so on, doubling each pile until the last square."

The rajah thought Chandra was foolish.

? How did Chandra trick the rajah? When will she have gathered more than 1 000 000 grains of rice?

Chandra's Spreadsheet

	B3		▼	=	=2*B2
	A		B	C	
1	Square on number of grains of rice				
2		1	2		
3		2	4		
4		3	8		
5		4	16		
6		5	32		

* A symbol used to represent multiplication in a spreadsheet

A. Cell B3 has the formula "=2*B2". Why does it make sense to use this formula here?

B. What would the formula for cell B6 be?

C. Predict the number of grains of rice on square 8. Calculate the number of grains on square 8. How close was your prediction?

D. Add a third column to the spreadsheet with the title Total grains of rice. What is the total for two squares? three squares? What formula can you use for the cells in this column?

E. Examine your spreadsheet. How do the numbers in the third column compare to the numbers in the second column?

F. Predict the number of chessboard squares covered when Chandra gets 1 000 000 grains of rice. Extend your spreadsheet to identify the number of squares. How close was your prediction?

Reflecting

1. a) Describe the pattern in the number of grains of rice.
   b) Why was Chandra able to use this pattern to trick the rajah?

Curious Math

Rice on a Chessboard

You will need
- rice
- different containers
- spreadsheet software

If the rajah gave Chandra all of her rice, how would she carry it away?

1 What is the smallest container you could use to hold all the grains of rice on the 1st row of Chandra's chessboard?

2 What is the smallest container that would hold all the rice in the 1st row and the next square? and the next square? and the next?

3 How many of the rajah's squares could you cover using one bag of rice?

4 Estimate how many squares of rice it would take to fill each item.
  a) a backpack
  b) your classroom
  c) your school

5 What could you use to contain all the rice on all 64 squares?

CHAPTER 1

6 Solve a Simpler Problem

You will need
- a calculator

Goal Solve problems by using a simpler problem.

It's a new year. You have some new faces in your class. You set up a series of introductions. Each person in your class will shake hands with everyone else once.

? How many handshakes will take place?

Raven's Solution

Understand the Problem
There are 29 students in the class. There will be too many handshakes to count.

Make a Plan
I'll use a plan with fewer people and build up to look for a pattern. I can determine the number of handshakes between three people, then four people, then five and so on.

I'll draw pictures and make a table. I will show myself as R and other people as A, B, C, and so on.

Number of Students	Actions	Handshakes	Drawing
2	R shakes with A	1	R —— A
3	R and A R and B A and B	2 + 1 = 3	R —— A | / B
4	R and A R and B R and C A and B A and C B and C	3 + 2 + 1 = 6	R —— A | X | B —— C
5	R and A R and B R and C R and D A and B A and C A and D B and C B and D C and D	4 + 3 + 2 + 1 = 10	R —— A |X| B C \ / D

Carry Out the Plan
I made the table.

With each new student, the number of handshakes increased by one less than the total number of students.

I predict that for six people there will be
5 + 4 + 3 + 2 + 1 = 15 handshakes.

So, with 29 students, there will be
28 + 27 + 26 + ... + 3 + 2 + 1 = 406 handshakes.

Look Back
I tested my model with five other students, and the 6 of us did shake hands 15 times, which confirms my prediction. There doesn't seem to be any reason why the pattern should not continue as it does.

Reflecting

1. How did Raven make the problem simpler?
2. Why did Raven think that her model would work for 29 people?

Checking

3. How many yellow tiles are in design 7?
    a) Make a plan to solve this problem.
    b) Carry out your plan. How many yellow tiles are in design 7?
    c) How many yellow tiles are in design 15?

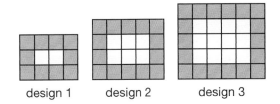

design 1   design 2   design 3

Practising

4. For the problem in Question 3, how many green tiles are in designs 7 and 15? Show your work.

5. Mya is using tiles to build a patio 20 tiles wide on each of two sides but in the shape of a triangle. How many tiles will she need? Show your work.

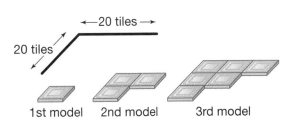

1st model   2nd model   3rd model

CHAPTER 1

7 Equal Expressions

Goal: Write equal expressions and determine the value of a missing term in an equation.

Rodrigo has three short-sleeved shirts and five long-sleeved shirts. Kurt has the same number of shirts. Kurt has four short-sleeved shirts.

? How many long-sleeved shirts does Kurt have?

Rodrigo's Solution

I have three short-sleeved shirts and five long-sleeved shirts.

I can represent the number of my shirts with the **expression** 3 + 5.

I know Kurt has four short-sleeved shirts. I don't know how many long-sleeved shirts he has. I can represent the number of Kurt's shirts with the expression 4 + ■.

I know that 3 + 5 = 8. So, it must be that 4 + ■ = 8. That means that ■ is 4. Kurt has four long-sleeved shirts. An expression for Kurt's shirts is 4 + 4.

Both expressions have the same value. I can write the **equation** 3 + 5 = 4 + 4. This equation means Kurt and I have the same number of shirts.

Communication Tip
The equal sign (=) says that the expression on the left is equal to the expression on the right. The two expressions are balanced, like masses on the pans of a balance beam.

3 + 5 = 4 + 4

Reflecting

1. In Rodrigo's expression 3 + 5, what does the 3 represent? What does the 5 represent?
2. Why did Rodrigo say that 4 + ■ = 8 and that ■ must be 4?
3. Rodrigo has four pairs of short pants and 5 pairs of long pants. Can you use the expression for Rodrigo's shirts and an expression for Rodrigo's pants to write an equation? Explain.

Checking

4. Each person is eating one fruit.
   a) Write an expression for the number of children and adults.
   b) Write an expression for the number of apples and oranges.
   c) Write an equation to show that the number of people and the number of fruit are equal.

Practising

5. Your class is going camping. Each student can bring one piece of luggage.
   a) Write an expression for the number of boys and girls.
   b) Write an expression for the number of knapsacks and suitcases.
   c) Is the number of pieces of luggage equal to the number of students? If so, use the expressions from parts a) and b) to write an equation. If not, explain why.

6. One day, 4 girls and 8 boys are playing musical chairs. There are 3 stools and 8 chairs.
   a) Write an expression for the number of children.
   b) Write an expression for the number of seats.
   c) If possible, write an equation with your expressions. If it is not possible, change the situation so that you can write an equation. Explain what you did.

7. If the expressions are equal, replace the ■ with an equals sign. If they are not equal, change one expression to make them equal.
   a) $1 + 2$ ■ $0 + 3$ b) $2 + 6$ ■ 4×2 c) $4 + 5$ ■ $8 + 2$

8. Replace each ■ so the two expressions in the equations are equal.
   a) $3 + 2 =$ ■ $+ 1$ b) $7 - 3 = 2 +$ ■   c) $2 \times 2 =$ ■ $+ 1$

9. Create a situation that the expressions in each equation might describe.
   a) $4 + 3 = 5 + 2$ b) $6 + 4 = 8 + 2$

CHAPTER 1

8 Variables in Equations

Goal Solve equations including symbols representing variables.

Evan had no money. His Aunt Sally and Uncle Wally each gave Evan some money, so he now has $16.

His aunt and uncle wanted to be fair, so they also gave Evan's sister, Adele, the same gifts. Adele says she used to have the same amount of money that Uncle Wally gave her, but now she has $20.

? How much did Aunt Sally give to each child?

Maggie's Solution

I can write an equation to represent Evan's money.
Aunt Sally's gift + Uncle Wally's gift = $16

I choose these symbols: S to represent the amount Aunt Sally gave and W to represent the amount Uncle Wally gave.

Now I can rewrite the equation for Evan's money: $\qquad$ $S + W = 16$

I can write an equation to represent Adele's money.
Aunt Sally's gift + Uncle Wally's gift + what Adele already had = $20

But what Adele already had is equal to Uncle Wally's gift so I can also represent that amount with W.

Now I can rewrite the equation for Adele's money: $\qquad$ $S + W + W = 20$

I see $S + W$ in both equations.

I know $S + W = 16$, so I can rewrite the equation for Adele's money: $\quad S + W + W = 20$
I know $16 + 4 = 20$, so the value of W must be 4. $\qquad\qquad\qquad 16 + W = 20$
This means that Uncle Wally gave $4. $\qquad\qquad\qquad\qquad\qquad W = 4$

To determine what Aunt Sally gave, I can rewrite the equation for $\quad S + W = 16$
Evan's money by replacing the W with 4. $\qquad\qquad\qquad\qquad S + 4 = 16$

I know $12 + 4 = 16$, so the value of S must be 12. $\qquad\qquad\qquad S = 16$

Aunt Sally gave each child $12.

Reflecting

1. How did Maggie use the equation for Evan's money in the equation for Adele's money?

2. Why did you need to know that Uncle Wally gave Adele the same amount as she had to solve the problem?

Checking

3. A baseball manager reported on the numbers of players on two teams. She said:
   Team 1 has 16 players, some experienced and some novices. Team 2 has 12 players. Team 2 has the same number of novices as Team 1, but has only three experienced players.
   a) Explain what situation is represented by this equation: $E + N = 16$
   b) Explain what situation is represented by this equation: $N + 3 = 12$
   c) How many novices are on Team 2?
   d) How many experienced players are on Team 1?

Practising

4. In a collection of 50 cards, Sal has some valuable and some regular cards. He has 16 more valuable cards than regular cards.
   a) Explain what information is represented by this equation: $V + R = 50$
   b) Explain what information is represented by this equation: $V = R + 16$
   c) Explain what information is represented by this equation: $R + 16 + R = 50$
   d) How many of each kind of card does Sal have?

5. In a fruit bowl there are twice as many apples as oranges. Altogether there are 9 pieces of fruit.
   a) Explain what information is represented by this equation: $A + O = 9$
   b) Explain what information is represented by this equation: $A = O + O$
   c) How many of each kind of fruit is there?

6. The perimeter of an isosceles triangle is 39 cm. The longest side is 6 cm longer than the shortest side.
   a) Could the equations $L + L + S = 39$ and $L = S + 6$ represent the lengths of the sides? Explain.
   b) Could the equations $S + S + L = 39$ and $L = S + 6$ represent the lengths of the sides? Explain.
   c) Determine the side lengths for both a) and b).

CHAPTER 1

Skills Bank

LESSON 1

1. Kendra is selling cookies for her community hall. She sells one box for $2.50.
   a) Make a table of values to show the cost for one to seven boxes.
   b) Determine the cost of eight boxes. Use a pattern rule. Show your work.

2. Lucas is selling apples for his community group. He sells 1 apple for $0.75.
   a) Make a table of values to show the cost of one to four apples.
   b) Determine the cost of eight apples. Use a pattern rule. Show your work.
   c) Determine the cost of 76 apples. Use a pattern rule. Show your work.

3. Frozen orange juice comes in 355 mL cans. To make one batch, you add three cans of water.
   a) What is the total volume of one to five batches of orange juice?
   b) Determine the volume of 10 batches. Use a pattern rule. Show your work.

LESSON 2

4. a) Determine the term number and perimeter of each of these four shapes.
   b) Determine the perimeter of the 10th shape. Use a pattern rule. Show your work.

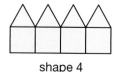

shape 1   shape 2   shape 3   shape 4

LESSON 3

5. There are four seats in each row of a bus except the last row, which has 5 seats. The total number of seats can be written as $5 + r \times 4$. Determine the number of seats in a bus for each value of r.
   a) 5   b) 12   c) 8

6. Six parents are going on a school trip with some students.
   a) Write a math expression describing the number of people on the trip. Represent the number of students with the symbol s.
   b) Determine the number of people on the trip for each value of s.
      i. 25   ii. 36

7. Julie is stacking nickels from her coin collection. There are five nickels in the first stack and each stack has one more nickel added to it.
   a) Graph the value of the stack in cents compared to the number of nickels.
   b) Describe the graph and the pattern.
   c) Determine the value of the nickels in the 10th stack using a pattern rule. Show your work.

8. a) Draw the next two shapes in this pattern.
   b) Record the number of squares in each of the 1st five shapes in a table.
   c) Graph the number of squares in each shape compared to the number of the shape.

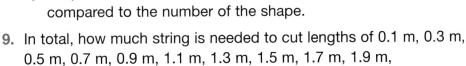

9. In total, how much string is needed to cut lengths of 0.1 m, 0.3 m, 0.5 m, 0.7 m, 0.9 m, 1.1 m, 1.3 m, 1.5 m, 1.7 m, 1.9 m, 2.1 m, 2.3 m, 2.5 m, 2.7 m, and 2.9 m?

10. In a 20 km bicycle race, there is a judge at the beginning, at the end, and at every kilometre in between. How many judges are there?

11. If the expressions are equal, replace the with an equals sign. If they are not equal, change a number in one expression to make them equal.
    a) 3 0 + 3
    b) 7 + 6 ■ 4 + 9
    c) 7 − 2 ■ 6 + 3
    d) 3 × 6 ■ 9 × 2
    e) 8 + 5 ■ 8 − 3
    f) 8 × 3 ■ 12 × 3

12. Write expressions for each situation. Then, if possible, write an equation for the situation. If it is not possible to write an equation, explain why.
    a)
    b)

CHAPTER 1

Problem Bank

1. Robyn is selling cookies. She sells one box for $2.50. For each additional box, up to 10 for the same customer, she can give a $0.10 discount on all the boxes. So one box costs $2.50. Two boxes cost $2.40 each (2 × $2.40) or $4.80.
   a) Determine the cost of one to five boxes for the same customer.
   b) Determine the total amount you would pay if you were to buy 10 boxes.

2. Sam is cooking soy sausage links for breakfast. He reads the directions: microwave two links for 1.5 min, four links for 2.5 min, and six links for 3.5 min.
   a) Determine how long Sam should cook one link.
   b) Determine how long Sam should cook 10 links.

3. Clarise is a fitness instructor in a gym. She is paid $50 each day, plus $10 for each fitness class she teaches.
   a) Determine the amount of money Clarise earns when she teaches from one to five classes in one day.
   b) Determine the amount of money Clarise would earn if she were to teach eight classes in one day.

4. A store sells boxed sets of a TV show on DVD. Each boxed set has six DVDs. The number of boxed sets the store sells is a variable.
   a) Write an expression describing the sales of DVDs using b to represent the number of boxes sold.
   b) Determine the number of DVDs sold for each value of b.
      i. 25   ii. 36

5. Noah collects $8 from each customer on his paper route.
   a) Determine the amount of money he collects from one to five customers.
   b) Graph the amount of money he collects compared to the number of customers he collects from.
   c) Determine how much money he collects from 30 customers using your graph.
   d) Noah buys his papers from a newspaper company for $200. From how many customers does Noah have to collect before he makes a profit?

6. In 2004, Lance Armstrong won the Tour de France bicycle race for the 6th time. He bicycled 3395 km at a speed of about 40 km each hour.
   a) Determine the distances Armstrong would ride from one to four hours.
   b) Determine the distance Armstrong would ride in 8 h.
   c) One stage of the race was about 240 km. About how long would it take Armstrong to bicycle this distance at his normal speed?
   d) What factors might affect a cyclist's speed? Does it make sense to use the average speed to predict times for different stages of a race? Is it a good mathematical model for the race?

7. Each term in this pattern is created by adding the two terms before it: 1, 1, 2, 3, 5, 8, … . What is the 20th term?

8. On Raz's 8th birthday his grandmother put $100 in a new bank account for him. She added $25 to the account 4 times a year and the bank added $2.50 per month. Determine how much money Raz will have on his 12th birthday.

9. Bicycle helmets are labelled on the top and the front of every box. If there are five rows of four boxes, with each new row stacked on top of the other row, how many labels are showing?

10. Phoenix is buying sports socks for the seven boys and eight girls on his team. The socks come in packages of three pairs. He writes this equation to describe the situation.

    $3 \times \blacksquare = 7 + 8$

    a) Explain what each expression in his equation represents.
    b) Determine how many packages of socks Phoenix should buy. Show your work.

11. Lana had a papaya, a kiwi, and a mango. She put them on the scale two at a time. She measured the mass of each pair of fruits as 140 g, 180 g, and 200 g. What was the mass of each fruit?

CHAPTER 1

Frequently Asked Questions

Q: How do spreadsheets use pattern rules to simplify calculations?

A: To use data from one cell to determine a value in another cell, a spreadsheet user inserts a pattern rule in the second cell. It tells the computer how to use the number from the first cell. For example, cell B2 is in column B and row 2. You can enter words, numbers, or formulas into a cell. For example, suppose a building has five windows on each floor. The spreadsheet calculates the number of windows for each number of floors. The formula in cell B2 is "=5*A2." It means multiply the value in cell A2 by 5.

	A	B
1	Floors	Windows
2	1	5
3	2	10
4	3	15
5		

Q: Why does an expression sometimes have only one value and sometimes more than one?

A: A mathematical expression might involve only numbers and operation signs. Then, it has only one value, for example, 5 + 8 has the value 13. But an expression might involve a variable. In that case, different values can be substituted for the variable and the expression takes on different values. For example, b × 3 has the value 3 when b = 1 but it has the value 6 when b = 2.

Q: What does the equals sign mean in an equation?

A: The equals sign shows that the expression on the left side is equal to the expression on the right side. It is like the balance point on a two-pan scale.

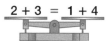

For example, the expression 2 + 3 is equal to 5.

Also, the expression 1 + 4 is equal to 5.

So you can write the equation 2 + 3 = 1 + 4.

Chapter Review

CHAPTER 1

1. One bicycle helmet costs $21.
   a) Determine the cost of one to five helmets. Use a table and show your work.
   b) Determine the cost of nine helmets. Use a pattern rule and show your work.

2. a) Draw the next two pictures in this pattern. Explain the pattern rule you used.
   b) Determine the number of squares in the 7th picture. Show your work.

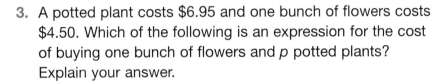

3. A potted plant costs $6.95 and one bunch of flowers costs $4.50. Which of the following is an expression for the cost of buying one bunch of flowers and p potted plants? Explain your answer.
   A. $6.95 + p \times 4.50$
   B. $6.95 + 4.50$
   C. $4.50 + p \times 6.95$

4. To enter a team in a badminton competition costs $20 plus $5 for each player.
   a) Determine the cost for one to four players to enter. Use a table of values and show your work.
   b) Graph the cost to enter compared to the number of players who enter. Determine the cost for six players to enter, using your graph.

5. On Sonja's 8th birthday her grandfather put $10 in a new bank account for her. Every year after that, he added $10 to the account twice a year. On her 12th birthday her grandfather gave Sonia $50 instead of $10. How much was in her account after her 12th birthday?

6. There are 8 teams in a bike relay race. Each team must race against each of the other teams once. How many races are needed?

7. If the expressions are equal, replace the ■ with an equals sign. If they are not equal, change one expression to make them equal.
   a) $7 + 3$ ■ $4 + 6$
   b) $9 - 2$ ■ $8 + 2$
   c) $10 + 23$ ■ 23×10

CHAPTER 1

Chapter Task

Patterns in Your Life

? **What kinds of patterns can you generate?**

A. Write two number patterns from your life. Pattern 1 should grow by a common difference. For example, "I sleep 8 h each night. One of my patterns is 8, 16, 24, 32, 40, 48, ..."

Pattern 2 should also grow, but not by a common difference. For example, "Each week I lift 10 kg. After three weeks, I increase the amount I lift by 2 kg, so my pattern is 10, 10, 10, 12, 12, 12, 14, ..."
Write at least six terms of each of your patterns.

B. Write a description of Pattern 1. Write two pattern rules for it.

C. Write a description of Pattern 2. Write a pattern rule for it.

D. Pose a problem about each pattern, such as "What is the 12th term in each pattern?" Use a table to solve one problem and a graph to solve the other one.

Task Checklist

☑ Did you use math language?

☑ Are your descriptions clear and organized?

☑ Did you show all your steps?

☑ Did you explain your thinking?

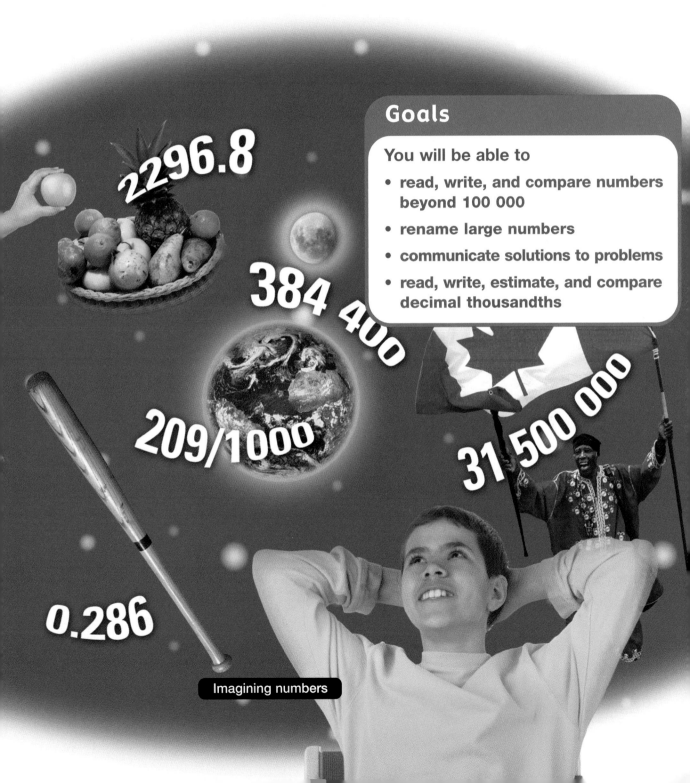

Numeration

CHAPTER 2

Goals

You will be able to
- read, write, and compare numbers beyond 100 000
- rename large numbers
- communicate solutions to problems
- read, write, estimate, and compare decimal thousandths

Imagining numbers

CHAPTER 2

Getting Started

> **You will need**
> - digit cards (0 to 9)
> - a place value chart
> - a 10-by-10 grid

Number Clues

Ayan and James are playing a number clues game. On Ayan's turn she arranges all of her digit cards from 0 to 9 into two numbers. She then gives James clues to help him arrange his digit cards into the same two numbers.

? How can you arrange all of the digits from 0 to 9 into two numbers to fit both sets of clues?

A. Ayan created these clues to describe her numbers. She used each of the 10 digits 0 to 9 once to make the two numbers.

First number	Second number
• It is between 50 000 and 52 000.	• It can be rounded to 30 000.
• There is a 2 in the tens place.	• There is a 9 in the ones place.
• The sum of the digits is 23.	

Name all the possible pairs of numbers to fit **both** clues at the same time.

B. What is the greatest possible value for the second number? How do you know?

C. Sketch or describe the base ten blocks you would need to represent the least possible second number.

D. Make up your own number with five different digits.

E. Make up a second number that uses the other digits that you didn't use in Part D.

F. Make up a set of clues that would allow someone to determine your numbers in Parts D and E. Include in your clue
- something about rounding
- something about a digit in a particular place

Do You Remember?

1. Write each number in words.
   a) 10 203   b) 37 123   c) 60 042

2. Write each number. Sketch or describe base ten blocks to model each number. Circle the model for the greater number.
   a) twenty-one thousand fourteen
   b) thirteen thousand two hundred five

3. 0.2 can be written "two tenths." Write each of these decimal numbers in words. Circle the words for the least number.
   a) 0.23   b) 0.4   c) 1.02

4. Write a decimal that is equivalent to 0.4. Draw a picture to represent it on a 10-by-10 grid.

5. Round each decimal to the nearest tenth.
   a) 0.23   b) 1.25   c) 3.89   d) 4.99

CHAPTER 2

1 Exploring Greater Numbers

You will need
- a calculator

Goal Compare numbers to one million.

Khaled did some research on the body for his science project. He found some interesting facts:

Eyes blink about 900 times an hour.
Hearts beat about 72 times a minute.
We take about 20 000 breaths a day.
The heart pumps 7200 L of blood a day.

To show how busy the body is, Khaled wants to estimate how long it takes to do each of these things **one million (1 000 000)** times.

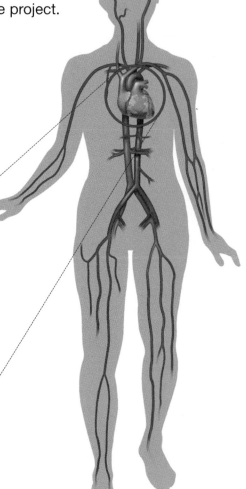

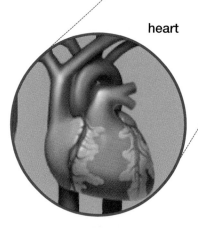

heart

? How long does it take to do something one million times?

A. Predict how long it takes for your heart to beat 1000 times.

B. Estimate the time in hours for your heart to beat one million times. Use the fact that one million is 1000 thousands.

one million
The number that is 1000 thousands
1 000 000

38

C. About how many days does it take for your heart to beat one million times?

D. About how long does it take to blink one million times? Show your work.

E. About how long does it take you to take one million breaths? Show your work.

F. About how long does it take your heart to pump 1 000 000 L of blood?

G. Research another body fact. Tell how it relates to the number one million.

Reflecting

1. What strategies did you use to answer Part A?
2. Describe any other strategies you used to answer other parts.
3. If someone asked you if one million is a lot, what would you say? Why?

Curious Math

Billions

In Canada, a billion is a thousand millions. In England, a billion is a million millions.

1. How would someone in each country write one billion in standard form?
2. What do you think a trillion is in each country?

CHAPTER 2

2 Reading and Writing Numbers

You will need
- a place value chart
- coloured counters

Goal Read, write, and describe numbers greater than 100 000.

Isabella read these facts in her almanac.

The distance from Earth to the Moon is about 384 400 km. The closest that Earth gets to Venus, our neighbour planet, is about 42 000 000 km.

Moon

Venus

? How do you read these distances?

Isabella's Models

I can model these numbers on a place value chart to help me read them.

The place value chart has **periods** of three. I'll model the numeral for the distance to the moon.
384 400 is in standard form.

Millions			Thousands			Ones		
Hundreds	Tens	Ones	Hundreds	Tens	Ones	Hundreds	Tens	Ones
			●● ●	●●● ●●●	●● ●●	●● ●●		
0 millions			384 thousands			400 ones		

I can read this number as "three hundred eighty-four thousand four hundred."

I can write it in expanded form as
300 thousands + 80 thousands + 4 thousands + 4 hundreds,
or 300 000 + 80 000 + 4000 + 400.

I model the shortest distance between Earth and Venus.

I see 42 in the millions period.
I can read this as "forty-two million."

Millions			Thousands			Ones		
Hundreds	Tens	Ones	Hundreds	Tens	Ones	Hundreds	Tens	Ones
	●● ●●	●●						
	42 millions			0 thousands			0 ones	

Communication Tip
We call a group of hundreds, tens, and ones a **period**. The first period has hundreds, tens, and ones. The second period has hundred thousands, ten thousands, and one thousands. The third period has hundred millions, ten millions, and one millions.

Reflecting

1. Do you think the distance to Venus is an estimate? Explain why or why not.
2. How does the place value chart show that one million is 1000 thousands?
3. How do the periods of three help you read numbers?

Checking

4. Isabella saw these numbers in her almanac.
   a) Describe the place value chart and counter model for 126 403. Write it in expanded form.
   b) Write the words for each number.

   126 403
   12 203 040
   537 146

5. Write each number as a numeral.
   a) three hundred thousand forty
   b) three million four thousand
   c) three hundred four million twenty thousand three

Practising

6. Write each number in standard and expanded form.

   a)

Millions			Thousands			Ones		
Hundreds	Tens	Ones	Hundreds	Tens	Ones	Hundreds	Tens	Ones

   b)

Millions			Thousands			Ones		
Hundreds	Tens	Ones	Hundreds	Tens	Ones	Hundreds	Tens	Ones

7. The distance from the planet Mercury to the Sun is about 57 million kilometres. Write that number in standard form.

8. a) Model three different six-digit numbers with eight counters on a place value chart. Describe each model.
   b) Write each number in standard and expanded form.

9. Write at least five numbers, in standard form, that include all of these words or phrases when they are read.
   two hundred   million   thousand   one hundred   forty
   One answer is 40 200 103.

Mercury

CHAPTER 2

3 Comparing and Ordering Numbers

You will need
- number lines

Goal Compare and order numbers to 1 000 000.

Rebecca is a hockey fan. She researched the total attendance at home games of Canadian hockey teams in 2003–2004.

Montreal Canadiens	Toronto Maple Leafs	Ottawa Senators	Calgary Flames	Vancouver Canucks	Edmonton Oilers
847 586	788 847	705 124	665 808	754 247	682 960

? Which team's attendance was between Edmonton's and Vancouver's?

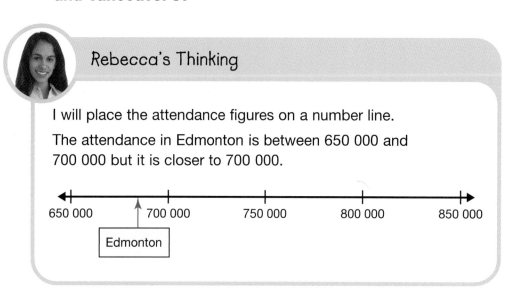

Rebecca's Thinking

I will place the attendance figures on a number line.

The attendance in Edmonton is between 650 000 and 700 000 but it is closer to 700 000.

650 000 — 700 000 (Edmonton) — 750 000 — 800 000 — 850 000

A. Place the attendance for each hockey team on a number line.

B. Compare the attendance for Edmonton and Calgary. Use an **inequality sign**.

C. Which teams had attendance between 700 000 and 800 000?

D. Which attendance is between Edmonton's and Vancouver's? How do you know?

Inequality sign
A sign to indicate that one quantity is greater than or less than another
5 > 2
12 000 < 13 000

Reflecting

1. How could you have eliminated Montreal as a possible answer to Part D without placing it on the number line?

2. Rodrigo says it's possible to compare the attendance numbers without putting them on a number line.
   Do you agree or disagree? Explain.

Checking

3. Rebecca also researched the total attendance for some women's basketball games.
   a) Order the attendance figures from least to greatest. Explain your strategy.
   b) The attendance in week 7 was between the attendances in weeks 3 and 5, but closer to that in week 3. Could it have been 121 000? Explain.

Weekly WNBA Attendance (2004)

Week	Total attendance
1	160 279
2	89 631
3	116 981
4	111 304
5	123 126
6	125 202

Practising

4. Michael's class is going to the Montreal Museum of Fine Arts. He researched attendance figures for some of their exhibits.
   a) Compare the attendance for the Picasso and da Vinci exhibits. Use an inequality sign.
   b) Order the attendance figures from least to greatest. Explain your strategy.

Exhibition	Attendance
Leonardo da Vinci: Engineer and Architect	436 419
Salvador Dali	192 963
Pablo Picasso: Meeting in Montreal	517 000
Monet at Giverny	256 738

5. a) Order these numbers from least to greatest.
      129 124, 75 212, 236 148, 124 113, 124 212, 750 212
   b) List three numbers between 124 113 and 129 124. Each number should have six different digits.

6. The number 3■4 146 is between 147 317 and ■22 367. The two missing digits are different. What might they be?

7. a) How many numbers are there between 100 000 and 125 000 where the digits are in order from least to greatest? No digit can be repeated.
   b) Which number is least and which is greatest?

CHAPTER 2

4 Renaming Numbers

You will need
- a place value chart

Goal Rename numbers using place value concepts.

An MP3 player uses compressed sound files so that they can be stored easily. Tom just got a new MP3 player. He is downloading songs for his sister.

The sizes of sound files are given in three different units:
- bytes
- kilobytes (kB, 1000 bytes)
- megabytes (MB, 1 000 000 bytes)

? Which song file will use the most space?

A Whole New World — 3.68 MB

Skip to My Lou — 233 848 bytes

Under The Sea — 1 431 430 bytes

Star Trek Theme — 1427.72 kB

I'm a Believer — 3132.5 kB

Tom's Comparisons

To compare the file sizes, I have to compare values using the same units. I could use bytes, kilobytes, or megabytes.
- I know that if the value is in bytes, the ones digit should be in the ones place of the place value chart.
- If the value is in kilobytes, the ones digit should be in the one thousands place of the chart.
- If the value is in megabytes, the ones digit should be in the one millions place of the chart.

I'll put in the missing zeros to write the full numbers.

A Whole New World will use the most space. 3.68 million bytes is 3 680 000 bytes. The next biggest file is 3132.5 thousand bytes. That's the same as 3 132 500 bytes, so it's less.

Millions			Thousands			Ones		
Hundreds	Tens	Ones	Hundreds	Tens	Ones	Hundreds	Tens	Ones
			2	3	3	8	4	8
		1	4	3	1	1	3	0
		3	6	8	0	0	0	0
		1	4	2	7	7	2	0
		3	1	3	2	5	0	0

44

Reflecting

1. Why could Tom have described the file sizes as 0.2 MB, 1.43 MB, 3.7 MB, 1.43 MB, and 3.1 MB to compare them?

2. Why did Tom put in zeros at the end of the numbers for the last three songs?

3. How else could you write the file sizes to compare them?

Checking

4. Three other music files were these sizes:
   1.43 MB   0.58 MB   3.12 MB
   a) Write each file size using byte measurements.
   b) Which file is less than 1 000 000 bytes?

5. Write each number in another form.
   a) 3.2 million = ■ ones
   b) 0.47 million = ■ thousands
   c) 0.32 million = ■ ones

Practising

6. Sarah downloaded some digital pictures. The file sizes were:
   285 kB   15 000 bytes   1.1 MB   371 245 bytes
   a) Write the first and third file sizes as a number of bytes.
   b) Estimate each file size except for the third one as millions of bytes, or megabytes.
   c) Which photo uses the most bytes?

7. Copy and complete Kayla's work:
   a) 3 145 276 is about ■ millions
   b) 3 145 276 is about ■ thousands
   c) 224 137 is about ■ millions

8. Explain why 0.1 million has to be 100 000.

9. Canadians ate 2296.8 million kilograms of fruit in 2002.
   a) About how many tonnes of fruit is this?
   b) There are about 30 million Canadians. Did a typical Canadian eat closer to 1 tonne, 0.1 tonnes, or 0.01 tonnes of fruit? Explain.

CHAPTER 2

5 Communicate About Solving Problems

You will need
- a calculator

Goal Explain your thinking when solving a problem.

Akeem read in the newspaper that the population of Canada increased from about 31.3 million in 2002 to about 31.5 million in 2003.

? How many additional Canadians were there each day?

Akeem's Drafts

I calculated how many additional Canadians there were each day.

I can give a better explanation of what I did.

My First Draft

The increase was about 0.2 million.

That's about 200 000.

The increase was about 550 people each day.

My Second Draft:

I wrote both values on a place value chart.

31.3 million is 31 300 000.

Millions			Thousands			Ones		
Hundreds	Tens	Ones	Hundreds	Tens	Ones	Hundreds	Tens	Ones
	3	1	3	0	0	0	0	0

31.5 million is 31 500 000.

Millions			Thousands			Ones		
Hundreds	Tens	Ones	Hundreds	Tens	Ones	Hundreds	Tens	Ones
	3	1	5	0	0	0	0	0

The difference is 200 000.

There are 365 days in a year. I divided 200 000 by 365 to calculate the number of additional people each day.

The newspaper said "about 31.3 million" and "about 31.5 million" so those numbers are rounded and not exact. The answer I got by dividing 200 000 by 365 is an estimate. It is about 550.

A. How is Akeem's first draft incomplete?

B. Describe how the second draft is better than the first draft.

Communication Checklist

- ☑ Did you explain your thinking?
- ☑ Did you use correct math language?
- ☑ Did you include enough detail?

Reflecting

1. How did Akeem's second draft show he used the Communication Checklist?

2. Could Akeem have improved the second draft more? Explain.

Checking

3. Akeem also read in the article that the population of Canada increased from about 29.7 million in 1996 to about 31.5 million in 2003. About how many additional Canadians were there each day?
   a) Write a first draft explaining the steps you used to solve this problem.
   b) How can you improve your first draft? Use the Communication Checklist. Write a second draft after a partner has had a look at it.

Practising

4. The number of personal computers in use in Canada increased from about 17.2 million in 2000 to 22.39 million in 2004. About how many additional computers were in use each day? Explain how you solved the problem.

5. Jacob found these population figures on the Internet. Compare the yearly and daily increase of population in Asia compared to Europe.

	1975	2000
Asia	2400 million	3700 million
Europe	676 million	727 million

   a) Write a first draft explaining the steps you used to solve this problem.
   b) How can you improve your first draft? Use the Communication Checklist. Then write a second draft.

CHAPTER 2

Frequently Asked Questions

Q: How do you read whole numbers with six or more digits?

A: A place value chart shows periods of three digits.

Within each period, you read the numbers in the same way.

The number modelled below is read "three hundred forty-two **million**, three hundred forty-two **thousand**, three hundred forty-two."
342 342 342

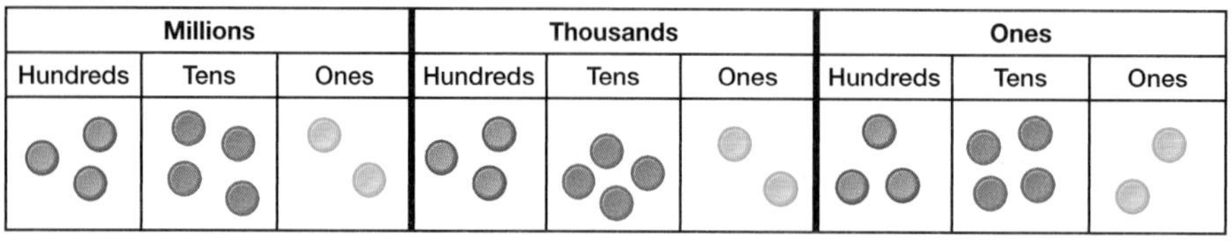

Millions			Thousands			Ones		
Hundreds	Tens	Ones	Hundreds	Tens	Ones	Hundreds	Tens	Ones
●●	●●●	●●	●●●	●●●●●	●●	●	●●●●	●●

The period to the left of the millions is the billions.

Q: How do you decide which of 2 six-digit numbers is greater?

A: Begin with the hundred thousands place. If one number has more hundred thousands, it is greater. If these digits are the same, compare the ten thousands digit to decide which is the greater one, and so on.

> These numbers are in order:
> 623 156
> 589 146
> 572 100
> 571 389
> 571 242

Q: How can you rename a whole number using place value ideas?

A: You use the relationship between adjacent place values to help you rename. For example, 452 130 can be modelled on a place value chart.

Millions			Thousands			Ones		
Hundreds	Tens	Ones	Hundreds	Tens	Ones	Hundreds	Tens	Ones
			4	5	2	1	3	0

Then you can see that it is
452.13 thousand
or about 0.45 million

CHAPTER 2

Mid-Chapter Review

1. How many of each unit makes one million?
   a) thousands   b) ten thousands   c) hundreds

2. About how many years are one million days?

3. Write each number in standard and expanded form.

   a)
Millions			Thousands			Ones		
Hundreds	Tens	Ones	Hundreds	Tens	Ones	Hundreds	Tens	Ones
			○	○○			○○ ○○	

   b)
Millions			Thousands			Ones		
Hundreds	Tens	Ones	Hundreds	Tens	Ones	Hundreds	Tens	Ones
			○○		○○○		○	○○○○

4. Write each number in words.
   a) 1 430 205   b) 3 126 100   c) 452 003

5. Christopher researched the area of some very large islands.
   a) Compare the sizes of Great Britain and Ireland. Use an inequality sign.
   b) Order the area of the islands from least to greatest area. Explain your strategy.

Island	Area in square kilometres
Ireland	84 426
Cuba	114 525
Baffin Island	476 068
Great Britain	229 883
Newfoundland	110 681

6. a) Describe 1 234 110 in at least two other ways using place value relationships.
   b) Choose one of those ways and describe why it's correct.

7. Copy and fill in the missing amount:
   a) 300 000 = ▬ million
   b) 30 000 = ▬ million
   c) 320 000 = ▬ million

8. The population of Ontario increased from about 10.4 million in 1991 to 12.2 million in 2003. About how many additional people were there each day? Explain your thinking.

CHAPTER 2

6 Reading and Writing Decimal Thousandths

You will need
- thousandths grids
- a decimal place value chart

Goal Read, write, and model decimals.

Marc is writing a report on the population of Canada. He found this table showing the number of people in each age group for every 1000 people in Canada in 2001.

Population by Age for Every 1000 People in 2001

Age group	Number for every 1000
Under 5 years	57
5–19 years	203
20–44 years	367
45–64 years	243
65 years or over	130

? How can a decimal be used to describe people who are 5 to 19 years old?

Marc's Explanation

I write a fraction for the 5 to 19 years old.

$\frac{203}{1000}$

I use a grid made up of 10 columns, each with 10 squares in it.

Each square is divided up into 10 rectangles that represent thousandths.

Each column is $\frac{100}{1000}$ or $\frac{1}{10}$.

Each square is $\frac{10}{1000}$ or $\frac{1}{100}$.

I colour 203 thousandths on a grid.

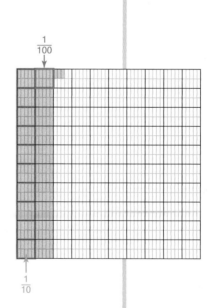

Next, I write the number on a place value chart.

Tens match tenths and hundreds match hundredths on opposite sides of the place value chart. Thousandths must be the next place to match thousands.

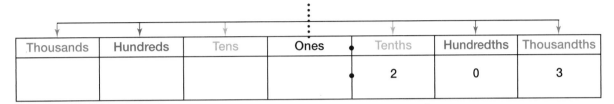

Thousands	Hundreds	Tens	Ones	Tenths	Hundredths	Thousandths
				2	0	3

In expanded form, 203 thousandths is
2 tenths + 0 hundredths + 3 thousandths.

The decimal for 5 to 19 year olds is 0.203.

Reflecting

1. How does the place value chart show that there are 10 thousandths in each hundredth (0.01) and 100 thousandths in each tenth (0.1)?

2. When might you read 0.203 as 2 tenths 3 thousandths? When might you read it as 203 thousandths?

Checking

3. Marc wants to represent each of the other age groups with a decimal as well.
   a) Write a fraction for each age group.
   b) Colour a thousandths grid for each age group.
   c) Write each number in expanded form.
   d) Write a decimal for each age group.

4. In words, 0.34 can be written as 34 hundredths. Write each decimal in words.
   a) 0.120   b) 0.007   c) 0.305

Practising

5. Which two age groups could be put together to cover more than 0.500 of the thousandths grid, or more than half of the total population?

6. The table shows the portion of Earth covered by three oceans.
   a) Write a fraction for each ocean.
   b) Colour a thousandths grid to represent each fraction.
   c) Write each number in expanded form.
   d) Write a decimal for each ocean.

Portion of Earth Covered by Oceans

Pacific Ocean	three hundred fifty-two thousandths
Atlantic Ocean	two hundred nine thousandths
Arctic Ocean	twenty-eight thousandths

7. Each of these distances is in thousandths of a metre.

   0.356 m
   0.892 m
   0.025 m

   a) Write the words you would say to read each decimal.
   b) Describe each distance in centimetres and millimetres.

8. Write a decimal in standard form to fit each description.
   a) one thousandth greater than 2.548
   b) one hundredth greater than 2.548
   c) one tenth greater than 2.548

9. a) List two fractions that are equivalent to 0.500.
   b) Rename 0.500 as a decimal hundredth and a decimal tenth. How do you know they are equivalent decimals?

10. Why is 0.455 halfway between 0.45 and 0.46?

11. A thousandths grid is completely coloured in when three different decimals are shown. One decimal is double one of the others.
    a) What could the three decimals be?
    b) How do you know there are many answers to part a)?

Math Game

Close as You Can

You will need
- digit cards

Number of players: 2 to 4

How to play: Arrange your cards into a number as close to the number dealt as possible.

Step 1 Shuffle the cards.
Deal out four cards to each player.

Step 2 Turn over the next four cards and place them face up in the order you deal them with a decimal point between the first and second cards.

Step 3 Each player arranges his or her cards to make a number as close to the face up numbers as possible.

Step 4 The player with the number closest to the face up number gets a point. The first person with 10 points wins.

1.573

2.245   1.634

Maggie's Hand

Emilio turned over 1.573.
He could make 2.245, but I could make 1.634.
I get a point since 1.634 is less than 0.1 away, but 2.245 is almost 1 away.

CHAPTER 2

7 Rounding Decimals

You will need
- thousandths grids
- number lines

Goal: Interpret rounded decimals and round decimals to the nearest tenth or hundredth.

Baseball players' batting averages are reported as decimal thousandths.

Li Ming had a little league batting average of 0.286. That means she can be expected to get 286 hits in 1000 times at bat.

? About how many hits would you expect Li Ming to get in 10 or 100 times at bat?

Li Ming's Solution

I model 0.286 on a thousandths grid.

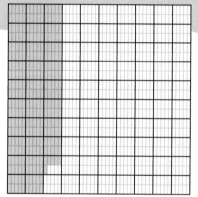

A. Li Ming coloured in 286 thousandths on the grid. About how many hundredths are shaded?

B. About how many hits in 100 times at bat would you expect Li Ming to get? Use your answer to Part A.

C. Round Li Ming's batting average to the nearest decimal hundredth.

D. On the thousandths grid, about how many tenths are shaded?

E. About how many hits in 10 times at bat would you expect Li Ming to get? Use your answer to Part D.

F. Round Li Ming's batting average to the nearest decimal tenth.

Reflecting

1. a) Describe how you would use a thousandths grid to round 0.718 to hundredths.
   b) Describe how you would use a thousandths grid to round 0.718 to tenths.

2. Nicholas says he could use this number line to round 0.286 to the nearest hundredth. Do you agree or disagree? Explain.

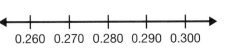

Checking

3. a) Model each baseball player's batting average on a thousandths grid.
   b) Round each number to the nearest hundredth.
   c) About how many hits in 100 times at bat would you expect each player to get?
   d) Round each batting average to the nearest tenth.
   e) About how many hits in 10 times at bat would you expect each player to get?

Player	Batting average
G. Zaun	0.367
R. Johnson	0.270
H. Clark	0.217
C. Woodward	0.235

Practising

4. Round each decimal to the nearest hundredth and to the nearest tenth.
   a) 0.158   b) 0.228   c) 0.067   d) 2.039

5. Which numbers below round to the same hundredth? Explain.
   0.234   0.324   0.237   0.229

6. Which decimal thousandths could be rounded correctly as described below? Mark all possible answers on a number line.
   a) up to 0.27 or up to 0.3   c) up to 0.28 or up to 0.3
   b) down to 0.25 or up to 0.3   d) up to 0.24 or down to 0.2

7. Andrew measured and cut a length of wood 1.32 m long. If he had measured to the nearest millimetre instead, what might the length of the wood be?

8. On a baseball team, the highest batting average is 0.338 and the lowest is 0.178. Would rounding to the nearest tenth be a good way to compare players? Explain.

CHAPTER 2

8 Comparing and Ordering Decimals

You will need
- a decimal place value chart
- number lines
- thousandths grid

Goal Compare and order decimals to thousandths.

Fast races, like the luge, are measured in thousandths of a second.
Moffat/Pothier's time in the 2002 Olympics men's double luge was 1:26.501.
That means 1 min and 26.501 s.

? Which pair of athletes won the race?

Men's Doubles Luge 2002 Olympics

Athletes	Time (s)
Grimmette/Martin	1:26.216
Thorpe/Ives	1:26.220
Moffat/Pothier	1:26.501
Leitner/Resch	1:26.082
Tchaban/Zikov	1:27.586

Tara's Thinking

If the decimal is greater, that means it took longer for the team to complete the run. I am looking for the least decimal.
All of the times are one minute and some seconds.
I only need to compare the seconds.
First, I will compare the times of Grimmette/Martin and Thorpe/Ives.

Grimmette/Martin

Tens	Ones	Tenths	Hundredths	Thousandths
2	6	2	1	6

Thorpe/Ives

Tens	Ones	Tenths	Hundredths	Thousandths
2	6	2	2	0

The tens, ones, and tenths digits are all the same.
I need to compare the hundredths digits.
1 hundredth is less than 2 hundredths, so
1:26.216 < 1:26.220.

A. Order the times from least to greatest.

B. Which pair of athletes won the race?

Reflecting

1. Why do you think the times for the luge are measured to the nearest thousandth of a second?
2. What strategy did you use to compare the times?
3. Joseph says he can compare decimal thousandths using a number line or a thousandths grid. Do you agree or disagree? Explain.

Checking

4. Some of the 2002 results of the women's single luge are shown.
   a) Compare the times of Neuner and Wilczak. Use an inequality sign.
   b) Order the times from least to greatest. Explain your strategy.
   c) Who won the race?

Women's Singles Luge 2002 Olympics

Athlete	Time (s)
Neuner	2:54.162
Wilczak	2:54.254
Niedernhuber	2:52.785
Otto	2:52.464
Kraushaar	2:52.865

Practising

5. Order these times from least to greatest.
   3:21.175   4:1.122   3:12.987   3:14.5

6. Which package of salmon fillets has the greatest mass?

7. Which is the greater distance, 2.198 km or 1315 m? Explain.

8. a) Order these decima[ls]
      1.024   0.305
   b) Change one digit in
      the numbers revers[ed]

9. Explain why 2.▪▪▪ is
   what numbers go in the

10. List the numbers of the
    3.8 that are greater tha[n]

11. Brittany bought three [packages for her]
    group's barbecue. The
    The mass of another w[as]
    package was in-betwee[n]
    package. List 3 possib[le]

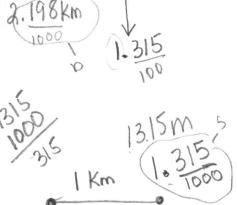

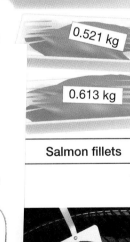

Salmon fillets

Mental Math

Dividing Decimals by Renaming

You will need
- a decimal place value chart
- counters

You can divide a decimal number by a whole number by renaming the decimal number.

To divide 2.0 by 5, represent 2 as 2 ones in a place value chart.

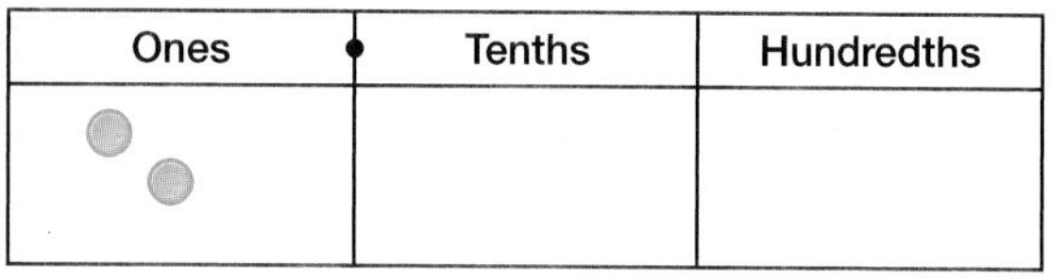

Rename 2 as 20 tenths.

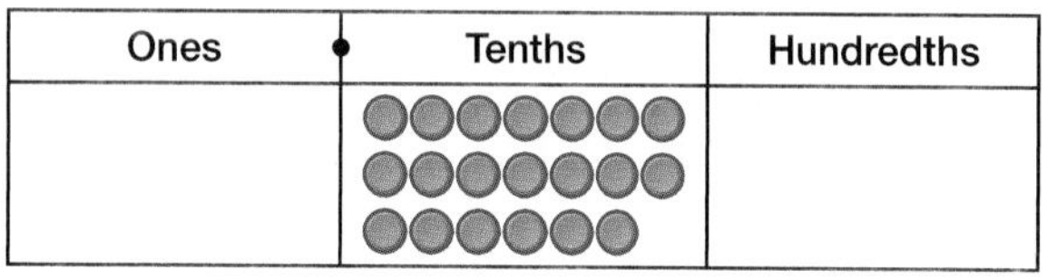

Now divide 20 tenths by 5.

$$5\overline{)20}\text{ tenths} = 4\text{ tenths} \qquad 5\overline{)2.0} = 0.4$$

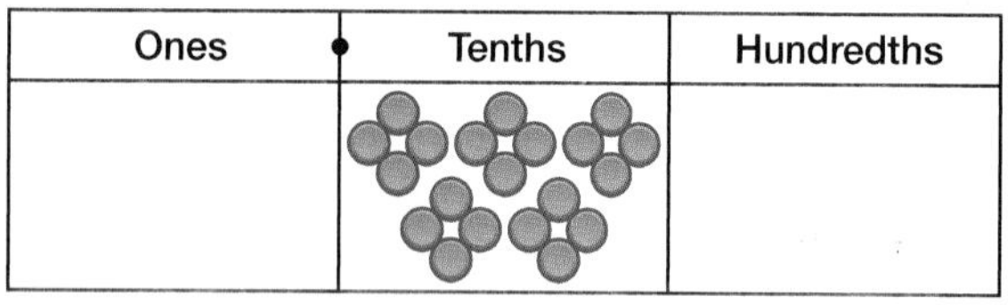

A. Why is 2 ones equal to 20 tenths?

Try These

1. Rename each decimal. Use mental math to divide.
   a) $1.0 \div 5$
   b) $6\overline{)1.8}$
   c) $4.0 \div 4$
   d) $9\overline{)6.3}$
   e) $3.5 \div 7$
   f) $8\overline{)4.0}$

CHAPTER 2

Skills Bank

LESSON

1
1. Copy and complete each
   a) 1 million = ■ thousa
   b) 1 million = ■ ten tho
   c) 1 million = ■ hundre

2
2. Write each number in wo
   a) 214 135   c) 1 2
   b) 300 002   d) 90

3. Write each number in sta
   a) six hundred sixty-five
   b) three hundred two tho
   c) four hundred fifteen t
   d) nine hundred seventy

4. Write each number in standard and expanded form.

   a)

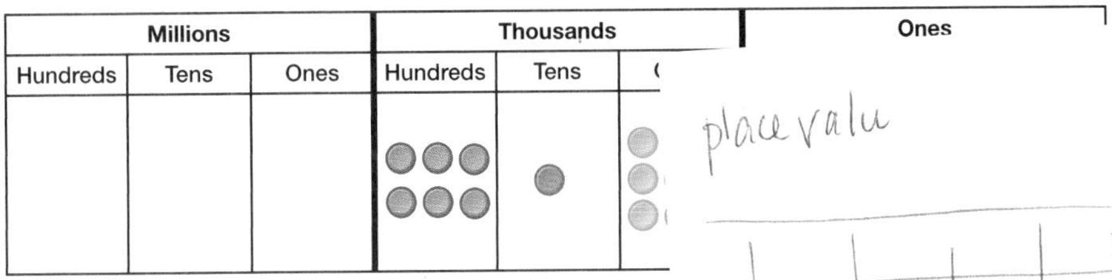

Millions			Thousands			Ones		
Hundreds	Tens	Ones	Hundreds	Tens				
			●●● ●●●	●				

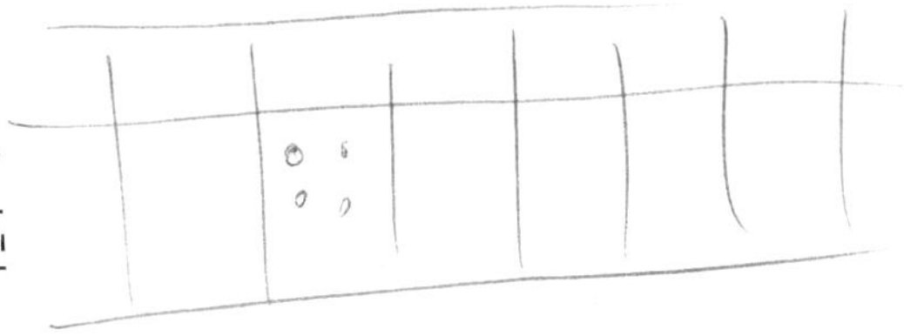

   b)

Millions			Thousands		
Hundreds	Tens	Ones	Hundreds	Tens	O
			●	●● ●	

5. Describe the place value chart and counter
   each number.
   a) 210 802   c) 311 022
   b) 570 089   d) 210 300

3
6. Copy and complete each number sentence using > or <.
   a) 189 673 ■ 189 771
   b) 601 176 ■ 623 489
   c) 797 881 ■ 797 781
   d) 598 329 ■ 598 392

(handwritten notes:)

standard 619 242

ex 600 000 + 10 000
    + 9000 + 200 + 40 + 2

place value

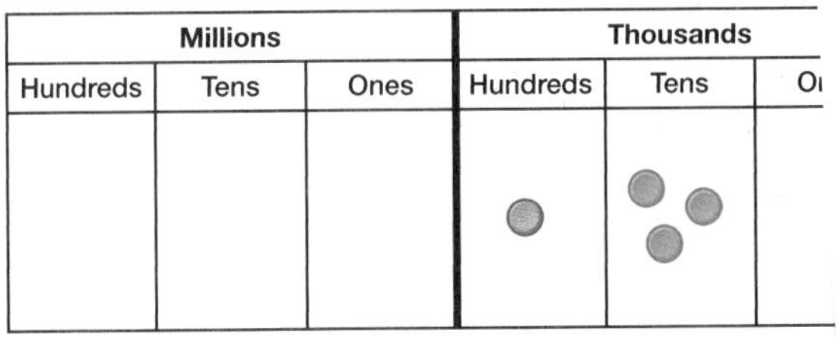

model

put circles in
the right place

7. Order each set of numbers from least to greatest.
   a) 264 135   234 510   310 007   300 176
   b) 410 207   123 126   57 103   57 301
   c) 200 020   220 002   221 121   212 121

8. Model each number on a place value chart.
   a) 1.235 million
   b) 0.568 million
   c) 212.3 thousand
   d) 1438.56 thousand
   e) 0.46 thousand

Millions			Thousands			Ones		
Hundreds	Tens	Ones	Hundreds	Tens	Ones	Hundreds	Tens	Ones

9. Copy and complete each equation.
   a) 4.2 million = _____ ones
   b) 0.57 million = _____ ones
   c) 212.3 thousand = _____ ones
   d) 600 000 = _____ millions
   e) 147 thousand = _____ millions

10. Write each fraction as a decimal thousandth.
    a) $\frac{246}{1000}$ c) $\frac{98}{1000}$ e) $\frac{21}{100}$
    b) $\frac{400}{1000}$ d) $\frac{9}{1000}$ f) $\frac{7}{10}$

11. Write each decimal in expanded form.
    a) 0.493   c) 0.206   e) 1.222   g) 0.005
    b) 0.527   d) 0.609   f) 4.030   h) 2.002

12. Write each decimal in words.
    a) 0.214   c) 0.207   e) 3.063   g) 0.003
    b) 0.100   d) 1.214   f) 2.091   h) 1.007

13. Model each decimal on a place value chart.

a)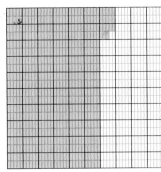

c)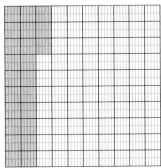

b)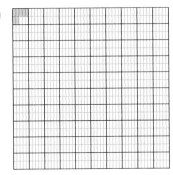

14. Model each decimal on a 1000ths grid.
    a) one hundred twenty-two thousandths
    b) seventy-five thousandths
    c) six thousandths

15. Round each decimal to the nearest hundredth.
    a) 0.216
    b) 0.556
    c) 1.312
    d) 2.004
    e) 3.007
    f) 2.939
    g) 2.635
    h) 1.996

16. Round each decimal to the nearest tenth.
    a) 3.418
    b) 1.572
    c) 8.007
    d) 0.001

17. Compare each pair of decimals. Use an inequality symbol.
    a) 0.931 ■ 0.831
    b) 3.543 ■ 3.354
    c) 1.111 ■ 1.110
    d) 0.078 ■ 0.009
    e) 0.590 ■ 0.591
    f) 0.43 ■ 0.198

18. Order each set of numbers from least to greatest.
    a) 6.092   6.187   5.989   6.989
    b) 11.756 10.981 10.643 11.173
    c) 3.120   31.25   0.204   0.73
    d) 0.023   3.002   3.020   0.302

19. List three decimals between 0.12 and 0.14.

CHAPTER 2

Problem Bank

LESSON

1
1. About how many individual boxes of raisins would you need to eat to get to one million raisins?

2. Do you think you have ever read a book with one million words? About how many pages would a book with one million words be?

2
3. About how many minutes old were you on your second birthday? Write your answer in the words you would say to read it.

4. Qi is reading a number. Some of the words he said to read it are shown; he might say other words as well. List six possible numbers he might have been reading.

3
5. How many numbers from 100 000 to 112 000 contain at least one zero?

6. What is the greatest six-digit even number that matches these clues?
   - The ten thousands digit is twice the tens digit.
   - The hundred thousands digit is more than six.
   - The thousands digit is divisible by the ones digit.
   - No digit is used more than once.

5
7. Rachael types 85 words each minute. She has typed 1 000 000 words. If she types for 6 hours each day, for about how many days has she typed? Explain the steps you used to solve this problem.

6
8. Write each of these measurements using whole numbers of units.
   a) 2.134 m   b) 123.25 km   c) 1.415 L

7
9. When you round a certain decimal less than one to the nearest tenth, it increases by 37 thousandths. What do you know about the decimal?

8
10. Show six ways to use each of the digits from 0 to 4 to make this inequality true.

CHAPTER 2

Frequently Asked Questions

Q: What does a decimal with three places, such as 0.015, mean?

A: The place to the right of the hundredths place is the thousandths. This is because ten thousandths, or $\frac{10}{1000}$, makes one hundredth, or $\frac{1}{100}$.

The number 0.015 could be read either as one hundredth $\left(\frac{1}{100}\text{ or }0.01\right)$ and five thousandths $\left(\frac{5}{1000}\text{ or }0.005\right)$ or as 15 thousandths $\left(\frac{15}{1000}\right)$.

Ones	•	Tenths	Hundredths	Thousandths
			●	● ● ● ●

Q: How do you round a decimal thousandth to the nearest tenth or hundredth?

A: You can use a thousandths grid to see which hundredth or tenth the amount is closest to.

0.248 is almost 25 hundredths, so it can be rounded to 0.25, and is closer to two tenths than three tenths, so it can be rounded to 0.2.

0.248 is closer to 0.25 than 0.24 and closer to 0.2 than 0.3 on a number line.

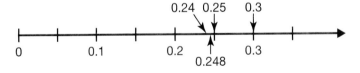

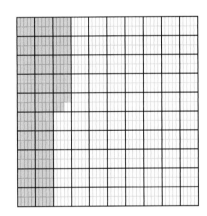

Q: How do you decide which of two decimals is greater?

A: If a decimal is farther to the right on a number line, it is greater. You can predict by starting at the digit that represents the greatest place value.

1.47 > 0.36 since 1 > 0
0.25 > 0.178 since 0.2 > 0.1

As soon as one number has a greater digit in that position, it is greater. If not, keep moving right until one digit is greater than another.

Chapter Review

CHAPTER 2

LESSON 1

1. Nicole can flex her fingers 15 times every 10 s. How long would it take her to flex her fingers about one million times?

LESSON 2

2. a) List three numbers between 100 000 and 400 000 that each use six different digits.
   b) Model each number on a place value chart. Sketch or describe each model.
   c) Write each number in expanded form.

3. Write each number in words.
   a) 3 120 048   b) 500 005   c) 12 316 148

LESSON 3

4. A number between 100 000 and 500 000 is greater than a number between 200 000 and 400 000. What do you know about the first number?

5. Brandon researched the areas of some large deserts.
   a) Compare the areas of the Chang Tang Desert and the Peruvian Desert. Use an inequality sign.
   b) Order the area of the deserts from least to greatest. Explain your strategy.

Desert	Area in square kilometres
Chang Tang Desert	800 310
Peruvian Desert	253 820
Great Sandy Desert	388 500
Namib Desert	207 977

LESSON 4

6. When you write the number 21.3 million, the left digit is in the ten millions column on the place value chart shown below.

Millions			Thousands			Ones		
Hundreds	Tens	Ones	Hundreds	Tens	Ones	Hundreds	Tens	Ones
	2	1	3	0	0	0	0	0

In which column is the leftmost digit for each of these numbers?
a) 3.2 million
b) 0.57 million
c) 0.03 million
d) 58.1 thousand

7. A digital photo uses more than 157 kB of space, but less than 0.2 MB. What might its size be?

8. The population of the Toronto area increased from about 4.3 million in 1996 to 4.7 million in 2001. About how many additional people were there each day in the Toronto area? Explain your thinking.

9. The Aboriginal population per 1000 people in four cities is shown in the chart.
   a) Write a fraction for the number of Aboriginal people in each city.
   b) Colour a thousandths grid for each number.
   c) Write each number in expanded form.
   d) Write each number in decimal form.

   Aboriginal Population for Every 1000 People

City	Number for every 1000
Kamloops, BC	64
Ottawa, ON	13
Regina, SK	83
Prince Albert SK	292

10. Write each as a decimal.
    a) three thousandths
    b) two hundred five thousandths
    c) thirty-six thousandths

11. Write the decimal thousandths for each position on the number line.

12. A recent survey showed that 217 out of every 1000 online shoppers in Canada are female.
    a) Write this as a decimal.
    b) Round the decimal to the nearest hundredth. Show your strategy.
    c) About how many shoppers out of 100 would be female?

13. Put the same digit in each blank so the numbers are in order from greatest to least.
    ■.100   3.1■2   3.18■   0.■33

CHAPTER 2

Chapter Task

Reporting Numbers

Pretend you are a TV newscaster. You have been given this news report to read on the air. You have circled the errors.

? **How can you use numbers effectively in a news report?**

A. Explain how you know that these are errors.

B. Write your own news report on a topic of interest to you. Include a variety of numbers and comparisons of numbers. Be prepared to read it to the class.

C. For each number, tell why your choice of the type of number was the best for providing information to the reader.

Task Checklist

- ☑ Did you include numbers larger than 100 000 as well as decimal thousandths?
- ☑ Did you include a rounded number?
- ☑ Did you include comparisons of numbers?
- ☑ Did you choose your words and numbers effectively?

Manitoba is a Growing Concern.

The census results are in. Our population has grown to 1 150 038, about (a million and a half,) with over half of our population, (471 274,) in our capital, Winnipeg.

Our birth rate of 0.124 is much higher than our death rate, (0.9,) so we're growing! In fact, our overall population growth is a big 0.001.

And did you know that we're generous, too? A typical Manitoban's donation to charity last year was $882. That would mean over ($10 million) from Winnipegers.

CHAPTER 3

Data Management

Goals

You will be able to

- collect, organize, and analyze the results of a survey
- make bar graphs, line graphs, and scatter plots
- use mean and median to compare sets of data
- describe how changing the scale or interval size affects the appearance of a graph
- use data to make arguments

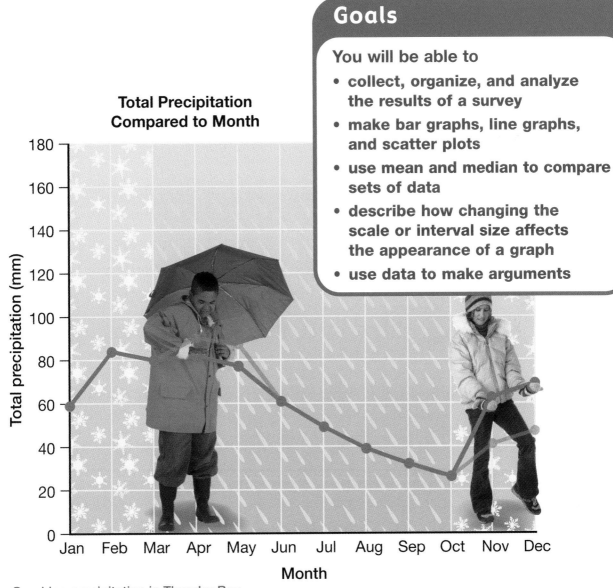

Graphing precipitation in Thunder Bay

CHAPTER 3

Getting Started

You will need
- grid paper or graphing software
- a ruler

Memorizing Pictures

Rudyard Kipling wrote several novels, including *The Jungle Book* and *Kim*. In *Kim*, Kim is shown a tray of precious stones for a period of time. The stones are covered with a cloth and he is asked to name all the stones he can remember.

? Does the memory of pictures improve with practice?

A. Study the pictures below for 1 min. Close your book, and then list as many pictures as you can remember on a sheet of paper. You have 1 min to complete your list.

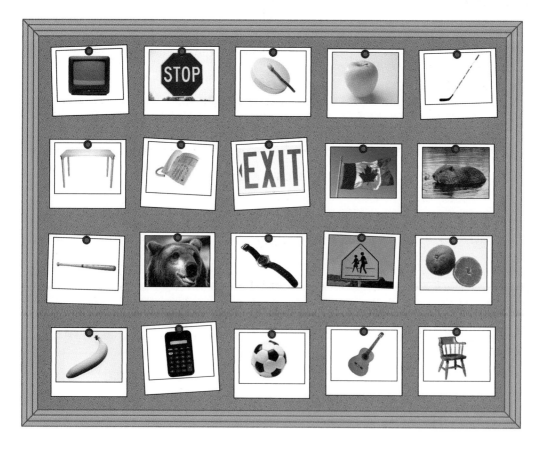

B. Count the number of pictures you remembered correctly. Record the number of pictures remembered by every student in the class in this first trial.

C. Repeat Parts A and B. Use a different sheet of paper to list the pictures for the second trial.

D. Organize your class data for both trials into intervals. Explain why you chose these intervals.

E. Create a double bar graph of the results of both trials. Explain how you chose your scale.

F. Does the memory of pictures improve with practice? Use the graph to justify your reasoning.

Do You Remember?

1. For a project on climate, Ethan researched the times of sunrise and sunset on the first day of each month in Edmonton. He used these times to calculate the number of daylight hours on those days and made a **broken-line graph**.

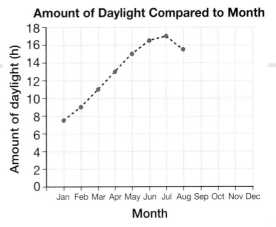

Month	Jan	Feb	Mar	Apr	May	Jun	Jul	Aug	Sep	Oct	Nov	Dec
Daylight hours	7.5	9.0	11.0	13.0	15.0	16.5	17.0	15.5	13.5	11.5	9.5	8.0

a) Copy and complete the graph using the data in the table.
b) Describe any trend you see in the graph.

2. Emma had 20 students in her class record the number of spam email messages they received one day.
   a) Organize her data in a tally chart with intervals.
   b) Emma plans to use the tally chart to make a bar graph with intervals. Describe how you think the bar graph will look.

20	6	25	45
15	20	8	26
35	63	4	16
15	12	22	6
10	12	43	31

CHAPTER 3

1 Creating and Analyzing a Survey

You will need
- grid paper
- graphing software

 Goal Collect, organize, and display the results of a survey.

Akeem wants to survey students in his school on how often they use the Internet in the evenings. He also wants to know how they use the Internet. These are two questions in his survey.

? **How can you organize and display the results of a survey?**

A. What other questions might Akeem include in his survey?

B. Conduct a survey about the Internet use of the students in your class. Use Akeem's questions or your own questions.

C. Show how to use a tally chart or spreadsheet to organize the results of one question.

D. How would you change your tally chart or spreadsheet if you were going to survey all of the students in your school instead of just your class?

E. Create a graph to display the results of one survey question.

F. Describe your graph.

G. Describe what the graph would look like if you used a different kind of graph.

1. How often do you use the Internet in the evenings?
   a) never
   b) a few times a month
   c) a few times a week
   d) almost every night
   e) every night

2. How often do you use the Internet for instant messaging?
   a) never
   b) a few times a month
   c) a few times a week
   d) almost every night
   e) every night

Reflecting

1. If you surveyed all of the students in your school, do you think that the results for your survey would be the same for each grade? Explain.

2. Explain why you chose the kind of graph you did for Part E.

3. Who might be interested in the results of your survey?

Mental Math

Determining Missing Decimals

Each decimal describes a part of the circle graph that is coloured.

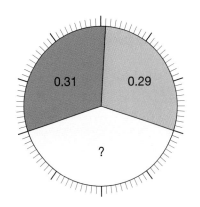

You can use mental math to determine the decimal that describes the yellow part.

0.31 + 0.29 = 0.60
1.00 − 0.60 = 0.40

A. How can you use mental math to add 0.31 and 0.29?

Try These

1. Determine the missing decimal. Use mental math.

   a) 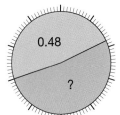   b)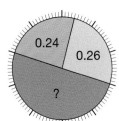

   c)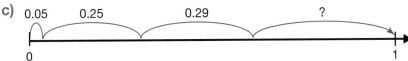

2. Determine the missing decimal.
   a) 1.0 − 0.1 = ■
   b) 0.65 + ■ = 1.00
   c) 0.75 + 0.05 + ■ = 1.00
   d) 2.50 + 4.5 + 0.75 = ■

71

CHAPTER 3

2 Plotting Coordinate Pairs

You will need
- grid paper
- ruler

Plot points on a grid and locate them using coordinate pairs.

Denise is playing a video game where her position is shown on a **coordinate grid**. She must move only left, right, up, or down on the lines of the grid.

The computer has given her a clue to find the door to the next level.

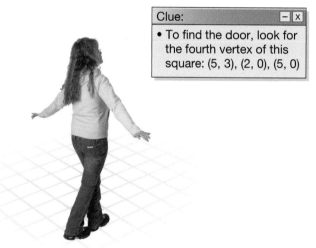

Clue:
- To find the door, look for the fourth vertex of this square: (5, 3), (2, 0), (5, 0).

? How can you use the clue to find the door?

coordinate grid
A grid with each horizontal and vertical line numbered in order

plot
Locate and draw a point on a coordinate grid

coordinate pair
A pair of numbers that describe a point where a vertical and a horizontal line meet on a coordinate grid

origin
The point on a coordinate grid at which the horizontal and vertical axes meet. The coordinate pair of the origin is (0, 0).

Denise's Reasoning

I need to **plot** the points (5, 3), (2, 0), and (5, 0).

In the **coordinate pair** for the point (5, 3), the first number, 5, describes the distance to the right of the **origin**. The second number, 3, describes the distance up from the origin.

To plot the point (5, 3), I first move 5 units to the right along the horizontal **axis**. Then I move 3 units upward along the grid line from 3. I draw the point and label it using its coordinate pair.

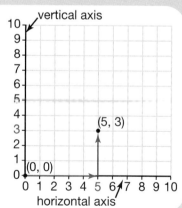

A. Create a coordinate grid on grid paper.
   Plot the points (5, 3), (2, 0), and (5, 0).

B. What does 0 in a coordinate pair mean in terms of movement?

C. Use the clue to locate the door. Plot and label the point.

Reflecting

1. Does the order of the two numbers in a coordinate pair matter? Use an example to help your explanation.

2. a) How can you tell without plotting that the coordinate pair (2, 0) is located on the horizontal axis?
   b) How can you tell without plotting that the coordinate pair (0, 2) is located on the vertical axis?

Checking

3. Denise can move to the next level in her game by solving this clue to find the door.
   Plot the point for the door on a coordinate grid.

   Clue:
   * To locate the door, go to the point midway between (3, 2) and (7, 2) on the grid.

Practising

4. a) Plot (3, 3) and (3, 6) on a grid. Connect them with a line.
   b) Describe the location of two other points on the line using coordinate pairs.

5. Each door in this grid is on a point. Describe the points using coordinate pairs.

6. a) Plot the point for each coordinate pair:
      (8, 6), (7, 5), (6, 4), (5, 3)
   b) What patterns do you notice about the numbers in the coordinate pairs?
   c) What pattern do you notice in the points plotted on the grid?
   d) Predict the coordinate pair of the next point in the pattern. Explain your reasoning.

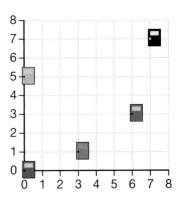

7. Make up your own problem about locating doors or other hidden objects on a grid. Give your problem to another student in your class to solve.

CHAPTER 3

3 Line Graphs

You will need
- grid paper or graphing software
- ruler

Goal Create and interpret line graphs.

During a class discussion of water use, Chandra wondered how much water her home wasted from a leaky outside tap.

? How can you use a line graph to predict how much water the tap wastes in a week?

Chandra's Line Graph

line graph
A graph of a line through points. It shows how the change in one value is related to change in another value.

Each day, I measured the number of litres of water that dripped from the tap into a bucket. I recorded the total amount of water wasted since I started.

I will make a **line graph** to compare the amount of water wasted to the number of days.

I'm comparing the amount of water wasted to the number of days, so I put days on the horizontal axis.

Leaky Tap Data

Days	1	2	3	4
Amount of water wasted (L)	3	6	9	12

I want to predict the amount of water after a week, so I need 7 days on the horizontal axis.

Over 4 days, the amount of water wasted was 12 L. So over 7 days, it might almost double to 24 L. I will use a scale from 0 to 24 on the vertical axis. Each unit will represent 2 L so the graph isn't too tall.

I make coordinate pairs using the data in the table. Days are on the horizontal axis, so the number of days is the first coordinate in the pair.

I plot the first two points at (1, 3) and (2, 6).

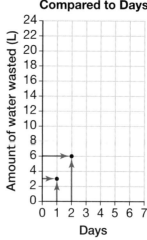

Amount of Water Wasted Compared to Days

A. Copy Chandra's line graph onto grid paper. Plot the remaining points and draw a line through the four points.
B. Describe the shape of the line graph.
C. Describe any patterns you see in each row of the table.
D. What rule can you use to calculate the amount of water wasted if you know the number of days?
E. Explain how to use the line graph to predict the number of litres of water wasted in 2.5 days.
F. Explain how to use the line graph to predict the number of litres of water lost in 5, 6, and 7 days.

Reflecting

1. What does the origin (0, 0) tell you about the amount of water wasted compared to the number of days?
2. How could you use patterns in the data table to predict the amount of water wasted in a week? Do your graph and the prediction from the table agree?
3. Explain how the shape of Chandra's line graph might change if the amount of water wasted each day were double the amount in her table.

Checking

4. Chandra found another leaky tap in her home. The table shows her recorded data.
   a) Use the data in the table to make a line graph.
   b) Explain how to use the table and graph to predict the amount of water wasted in 7 days.
   c) Explain how to use the line graph to predict the amount of water wasted in 2.5 days.

Leaky Tap Data

Days	1	2	3	4	5
Amount of water wasted (L)	4	8	12	16	20

Practising

5. Melvin has a remote-controlled truck. He wants to know the distance the truck can travel in 5 s. To find out, he tested the truck in a parking lot. Each second, he recorded the distance the truck had travelled.

    Remote Controlled Truck Data

Time (s)	1	2	3	4
Distance (m)	6	12	18	24

    a) Use the data in the table to create a line graph.
    b) Describe the shape of the line graph.
    c) Use the graph or table to predict the distance the car can travel in 5 s. Explain your reasoning.
    d) Use the graph or table to estimate the distance the truck can travel in 3.5 s.
    e) Compare the rows of the table. What rule can you use to calculate the distance if you know the time?

6. One line graph shows the distance travelled by Juan in his power wheelchair. The other line graph shows the distance walked by Lee in the same time. Which person travels a greater distance in the same time? Explain your reasoning.

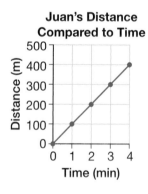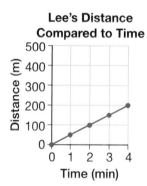

7. Matthew filled a 10 L container with water. Then he opened a tap at the bottom to empty it. The graph shows the amount of water in the container over time.
    a) Which part of the line graph shows the container being filled? Explain.
    b) Which part of the line graph shows the container being emptied? Explain.
    c) When did Matthew open the tap? Explain.
    d) Which took longer, filling or emptying? Explain.

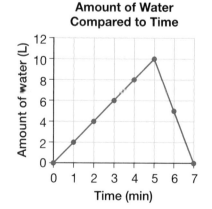

Curious Math

Telling Stories about Graphs

Marc is sketching a line graph to describe a hike he took in the mountains.

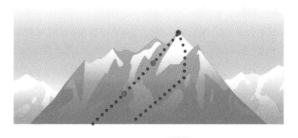

Marc's Hike

We climbed up to the summit and then back down again. We took three breaks during the hike, so there are three flat sections on the graph. The first two breaks were on the climb up, and the last one was at the summit.

1 Which graph matches Marc's description?

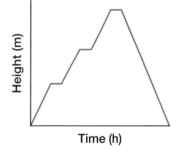

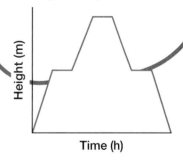

2 Why else might a graph describing a hike have flat sections?

3 Write a story or draw a picture that describes this graph of a hike.

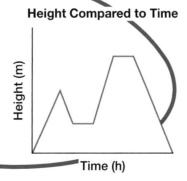

CHAPTER 3

4 Scatter Plots

You will need
- grid paper or graphing software
- a ruler

Goal Create and interpret scatter plots.

Qi examined the results of the 15 teams in the Eastern Conference of the National Hockey League (NHL) in a past season. He recorded the number of goals and wins in a table. He predicted that teams that score the most goals usually get the most wins.

NHL Eastern Conference Results

Team	Goals	Wins
Ottawa	262	43
Tampa Bay	245	46
Toronto	242	45
NY Islanders	237	38
Philadelphia	229	40
Buffalo	220	37
Atlanta	214	33
New Jersey	213	43
Boston	209	41
Montreal	208	41
NY Rangers	206	27
Pittsburgh	190	23
Florida	188	28
Washington	186	23
Carolina	172	28

? How can you compare the number of wins to the number of goals scored?

Qi's Scatter Plot

I'll make a **scatter plot** to organize and interpret the data.

Step 1 I'm comparing wins to goals, so I put goals on the horizontal axis.

All teams scored between 170 and 270 goals. I will use a scale of 1 unit to represent 10 goals.

The vertical axis shows the number of wins. All teams had between 20 and 50 wins. I will use a scale of 1 unit to represent 5 wins.

scatter plot

A graph made by plotting coordinate pairs to show if one set of data can be used to make predictions about another set of data

Step 2 I make coordinate pairs using the data in the table. The coordinate pair for the Ottawa point is (262, 43).
I estimate the location of the point (262, 43) and plot the point.

Step 3 I plot the points for the other 14 teams.

Step 4 I check my prediction by looking at the points on the scatter plot. It looks like the points go from the bottom left to the top right of the scatter plot.
I use my ruler to be sure. The points seem to follow an upward direction.

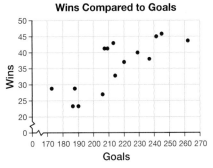

Step 5 The scatter plot shows that fewer goals usually result in fewer wins and more goals usually result in more wins.

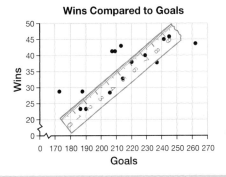

Reflecting

1. Why did Qi use the symbol ⌇ on each axis of the scatter plot?

2. Explain how to use the scatter plot to identify the team that scored the most goals.

3. How can you use Qi's scatter plot to predict the number of wins by a team that scores 200 goals?

4. How are a scatter plot and a line graph similar? How are they different?

Checking

5. The chart shows the number of goals and wins for the 15 Western Conference teams in the NHL.
   a) Compare wins to goals. Use a scatter plot.
   b) Do these teams usually get more wins if they score more goals? Use your scatter plot to justify your reasoning.

Practising

6. The chart shows the number of goals scored against and the number of points for the 16 teams in the first round in the Women's World Cup of soccer in 2003.
   a) Compare total points to goals against. Use a scatter plot.
   b) Describe how the points appear on the scatter plot.
   c) Compare how the points appear on this scatter plot with how the points appear in the hockey scatter plots. Describe any differences you notice.
   d) Do these teams usually get more points if they have fewer goals scored against? Use the scatter plot to justify your reasoning.

7. Isabel rolled a white die and a red die 50 times. She created a scatter plot comparing the numbers shown on the white die with the numbers shown on the red die.
   a) Compare how the points appear on this scatter plot with how the points appear on the other scatter plots you have made. Describe any differences you notice.
   b) Can you use the scatter plot to predict the number rolled on a red die if you know the number rolled on a white die? Use the scatter plot to justify your reasoning.

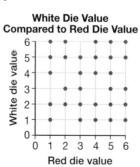

White Die Value Compared to Red Die Value

NHL Western Conference Results

Team	Goals	Wins
Detroit	255	48
Colorado	236	40
Vancouver	235	43
Edmonton	221	36
San Jose	219	43
Nashville	216	38
Los Angeles	205	28
Calgary	200	42
Dallas	194	41
St. Louis	191	39
Minnesota	188	30
Phoenix	188	22
Chicago	188	20
Anaheim	184	29
Columbus	177	25

Women's World Cup Results

Team	Goals Against	Total Points
U.S	1	9
Sweden	3	6
North Korea	4	3
Nigeria	11	0
Brazil	2	7
Norway	5	6
France	3	4
South Korea	11	0
Germany	2	9
Canada	5	6
Japan	6	3
Argentina	15	0
China	1	7
Russia	2	6
Ghana	3	3
Australia	5	1

Math Game

4 in a Row

You will need
- grid paper

Number of players: 2
How to play: play Tic-Tac-Toe on a grid.

Step 1   Player 1 writes down a coordinate pair in the X column. Player 1 draws an X to show the location of this coordinate pair on the grid.

Step 2   Player 2 writes down a different coordinate pair in the O column. Player 2 draws an O to show the location of this coordinate pair on a grid.

Step 3   Take turns placing points until a player gets either 4 X's or 4 O's in a row (up and down, across, or diagonally).

Raven's Turn

If I write (6, 5), I can win the game because I will have 4 Xs in a row.

X	O
(3, 3)	(3, 4)
(4, 3)	(5, 3)
(3, 2)	(3, 1)
(2, 3)	(1, 3)
(5, 4)	(2, 1)
(6, 5)	

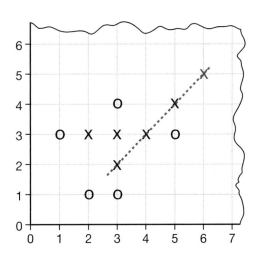

CHAPTER 3

Frequently Asked Questions

Q: How do I plot a point using its coordinate pair?

A: The first number in a coordinate pair shows the distance from the origin along the horizontal axis. The second number shows the distance from the origin along the vertical axis.

The grid shows the points (3, 5) and the origin, (0, 0). The coordinate pair (3, 5) describes a point that is 3 units along the horizontal axis and 5 units up the vertical axis from (0, 0).

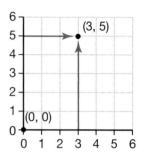

Q: Why do you use a line graph?

A: You can use a line graph to identify relationships between two sets of data.

The line graph shows coordinate pairs created from the data in the table. It can be used to predict values that were not collected. These values can be either along the line between points or along the line extending from the points.

Heartbeats Over Time

Time (min)	Heartbeats
0	0
1	80
2	160
3	240
4	320

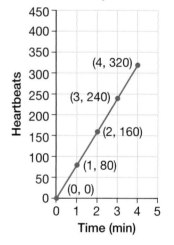

For example, after 1.5 min we would expect 120 heartbeats and after 5 min we would expect 400 heartbeats.

Q: What is a scatter plot?

A: A scatter plot is a graph showing points that represent two sets of data or measurements. The points are scattered and are not connected by lines.

For example, players of the online video game Soccer Shootout were asked to play with their stronger or dominant hand and then their opposite hand. The scores of each player describe a point on this scatter plot.

There seems to be a relationship. The scatter plot shows that players who score well with one hand also score well with the other hand.

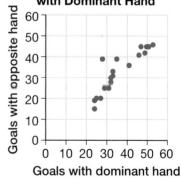

CHAPTER 3
Mid-Chapter Review

LESSON

1 1. Tiffany surveyed Grade 6 students in her school who like to play video games. The tally chart shows the results for one question.
   a) What type of graph would you use to display her data? Justify your choice.
   b) What is another question she might want to ask?
   c) Would the results of her survey likely apply to Grade 1 students? Explain.

Which game console do you prefer?
a) Play Station 2	|||| |||| |||
b) X-Box	|||| |||| ||||	
c) Game Cube	|||| |||| ||
d) Other	||||

2 2. List five coordinate pairs where the second number is one greater than the first number. Plot each coordinate pair on a grid. What do you notice about the points?

3 3. Ben is raising money for the food bank.
   a) Make a line graph comparing people fed to the amount of the donation.
   b) Explain how to use the graph to predict the donation that would feed 750 people.
   c) The food bank says that a donation of $187.50 would be enough to make 2.5 pots of soup. Use the table to show that this claim is reasonable.
   d) Use the graph to predict how many people can be fed with a donation of $187.50.

Food Bank Facts

Donation ($)	Pots of soup	People fed
0	0	0
75	1	150
150	2	300
225	3	450
300	4	600

4 4. The scatter plot shows the number of touchdowns against and points for the nine teams in the Canadian Football League.
   a) How many teams had more than 20 points?
   b) How many teams had fewer than 40 touchdowns scored against them?
   c) Describe how the points appear on the graph. Why would you expect the points on the graph to be in this direction?

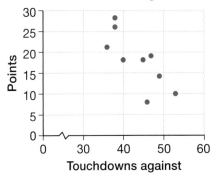

Points Compared to Touchdowns Against

CHAPTER 3

5 Mean and Median

You will need
- a calculator

Goal Use mean and median to compare sets of data.

A bird refuge is asking for donations to sponsor three snowy owls. The owl with the most support will have its picture posted on the refuge's Web site.

? Which owl has the most support?

Owl	Donations
Onohdo	$1, $2, $1, $1, $200
Gwaoh	$1, $2, $10, $1, $2, $10, $2, $5, $1, $2, $2, $5, $10, $2, $20
Ogra	$1, $1, $20, $20, $50, $10, $10, $5, $2, $1

Rebecca's Solution

I added and counted the donations. I recorded my work in a table.

Onohdo has the greatest total. Gwaoh has the most donations. I think that one of those two should be on the web site. I can use the **mean** to decide which one. I don't need to divide to know that Onohdo has a greater mean donation than Gwaoh. Onohdo should be on the web site.

	Total donation	Number of donations
Onohdo	$205	5
Gwaoh	$75	15
Ogra	$120	10

mean
A typical value for a set of numbers, determined by calculating the sum of the numbers and dividing by the number of numbers in the set

3, 4, 5, 2, 2, 3, 2
7 numbers in the set
the sum is 21
21 ÷ 7 = 3
The mean of this set is 3.

Li Ming's Solution

I think it's unfair to use the total donation or the mean donation. The total and the mean for Ohnodo are greater because of one big donation. I'll use the **median** to compare the donations.
I can determine the median by ordering the donations for each owl and identifying the middle number. First, I order the donation amounts for Onohdo.

$1 $1 $1 $2 $200

84

A. Determine the mean and median donations for each owl.
B. Which owls had the greatest mean and median donations?
C. Which owl do you think has the most support? Justify your choice.

> **median**
> A typical value for a set of numbers, determined by ordering the numbers and identifying the middle number
> 4, 5, 2, 3, 4, → 2, 3, 4, 4, 5
> The middle number or median of these five numbers is 4.
> 3, 2, 5, 6, 9, 11 → 2, 3, 5, 6, 9, 11
> 5.5
> The middle number, or median, of these six numbers is 5.5 because 5.5 is in the middle of 5 and 6.

Reflecting

1. Why did Rebecca not have to divide to know that Onohdo had a greater mean donation than Gwaoh?
2. Why did Li Ming order the numbers first before she determined the middle number?
3. Why can both the mean and the median be thought of as a single number that represents a set of numbers? Use an example.

Checking

4. The chart shows the donations given for each wolf at a zoo.
   a) Determine the mean and median donation for each wolf.
   b) Explain how you can determine the greatest mean without dividing.
   c) Which wolf do you think has the most support? Explain.

Wolf	Donations
Hado:wa:s	$1, $2, $1, $20, $5, $1, $10, $5, $2, $20, $5, $10, $10, $20
Hadri:yohs	$50, $50, $5, $1, $2, $2, $2
Mahikan	$2, $2, $1, $1, $2, $1, $2, $100

Practising

5. Identify the mean and the median of each set of numbers.
   a) 12, 1, 1, 2, 3, 4, 1, 5, 7, 7, 1
   b) 0, 6, 0, 3, 4, 12, 7, 0
   c) 1, 4, 3, 3, 0, 13
   d) 4, 4, 4, 4, 4

6. This chart shows the attendance at three pow-wows.
   a) Calculate the mean and the median daily attendance for each pow-wow.
   b) Which pow-wow had the best attendance? Justify your choice.
   c) Can you use the total attendance to determine which pow-wow had the best attendance? Explain.

Daily Attendance at Pow-Wows

6-day pow-wow	1200, 1600, 1300, 2000, 1800, 1100
4-day pow-wow	1500, 2000, 4000, 1500
3-day pow-wow	2500, 3500, 3000

7. Create two sets of numbers where the mean of one set is 5 greater than the mean of the other set. Explain your reasoning.

CHAPTER 3

6 Changing the Intervals on a Graph

You will need
- graphing software

Goal
Describe how changing the number of intervals changes a graph.

Maggie plays a board game. Each turn, she rolls two dice to determine the number of spaces to move. She wonders which sums are most likely to be tossed.

She uses an online simulator to roll two dice 1000 times. The table shows the number of times each sum was rolled.

1000 Dice Rolls

Sum	Rolls
2	39
3	50
4	63
5	118
6	132
7	190
8	146
9	121
10	75
11	43
12	23

? **How can you change the intervals in a graph to help you answer questions about the data?**

A. Enter Maggie's data into a spreadsheet or graphing software. Create a graph using the 11 sums. Justify your choice of graph.

B. What is the most likely number of spaces Maggie will move? Use your graph to justify your choice.

C. Maggie picked this card while playing her game. Predict the appearance of your graph if you use the two intervals 2 to 6 and 7 to 12. Make a graph to check your prediction.

> Roll 2 to 6: Free turn
> Roll 7 to 12: Lose a turn

D. How you can use your graph in Part C to determine whether the probability of getting a free turn is greater than the probability of losing a turn?

E. Maggie picked this card. Create a graph to determine which option is more likely.

> Roll 4 or 5: Gain 5 points
> Roll 8 or 9: Lose 5 points

F. Make up an event about a board game that uses the sum of two dice. Show how to use a graph to estimate the probability of that event happening.

Reflecting

1. Describe how the appearance of a graph with intervals changes for each situation.
   a) You increase the size of the intervals.
   b) You decrease the size of the intervals.

7 Changing the Scale on a Graph

You will need
- graphing software

Goal

Describe how changing the scale changes a line graph.

Kurt has been reading the science fiction book *20 000 Leagues Under the Sea* by Jules Verne.

He learned that 1 league is about 6 km. He created a table of leagues and kilometres. He used the data in the table to create a line graph.

Distance Comparison

Leagues	Kilometres
0	0
5000	30 000
10 000	60 000
15 000	90 000
20 000	120 000

? How will the appearance of the line graph change when the scale changes?

A. Enter Kurt's data into a spreadsheet or graphing software. Create a line graph with the same scale as Kurt's.

B. Predict the appearance of the line graph if you double the value of each unit on the scale of the vertical axis. Make a line graph to check your prediction.

C. Predict the appearance of the line graph if you double each value on the vertical scale again. Make a line graph to check your prediction.

D. Predict the appearance of the line graph if you increase or decrease the scale by a chosen amount. Make a line graph to check your prediction.

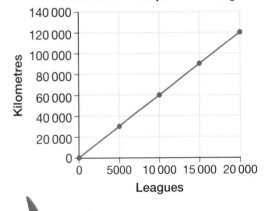

Reflecting

1. Describe how the appearance of a line graph changes for each situation.
   a) You increase the value of each unit on the scale on the vertical axis.
   b) You decrease the value of each unit on the scale on the vertical axis.

CHAPTER 3

8 Communicate about Conclusions from Data Displays

Goal Use data presented in tables, charts, and graphs to create an argument.

A newspaper gave a spelling test to students from Grades 1 to 6. The height of each student was also measured. The newspaper printed this headline and a scatter plot of the data.

Tom thinks the headline is unfair, and he decides to send an e-mail to the editor.

? What arguments can Tom make in his Letter to the Editor?

Tall people are better spellers than short people

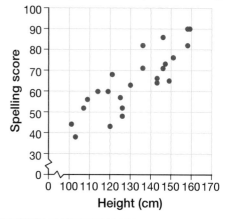

Tom's Letter to the Editor

I wrote a first draft.

> Dear Editor,
> Your graph shows that taller people are better spellers. But I don't think it's because they are taller. It's because you tested different ages of students.

I don't think I explained my idea well.

I wrote a second draft.

> Dear Editor,
> The points on the top right of the graph show that tall students have high spelling scores. The points on the bottom left show that short students have low spelling scores.
>
> But you used students from Grades 1 to 6. The short students are the younger students. The tall students are the older students. Older students are usually better spellers than younger students.
>
> I think the scatter plot really shows that older students are better spellers than younger students. Your study probably wouldn't show that tall students are better spellers if you tested only Grade 6 students.
>
> Tom

A. How did Tom improve his first draft? Use the Communication Checklist.

Communication Checklist

☑ Did you explain your thinking?
☑ Did you include enough detail?
☑ Did you use correct math language?
☑ Does your argument make sense?

Reflecting

1. What would a graph look like if it showed that Tom's argument is correct? Explain.

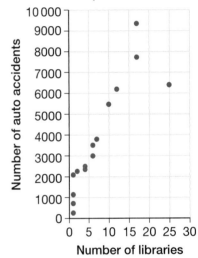

Checking

2. A newspaper reported a study on the relationship between the number of libraries in 20 Canadian cities and the number of auto accidents each year.
   a) Write a letter to the editor about the headline and the scatter plot. Explain what the scatter plot seems to show.
   b) Use the Communication Checklist to help you find ways to improve your letter.

Practising

3. Tanya counted the number of times she turned the pedals of her bike. She used her bike odometer to measure the distance her bike travelled. She made a line graph of the results.
   What conclusions can you draw from the line graph? Use the Communication Checklist to help you support your conclusion.

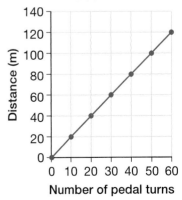

CHAPTER 3

9 Constructing Graphic Organizers

You will need
- software to draw graphic organizers

Goal Use Venn diagrams and Carroll diagrams to describe relationships between two sets of data.

The students in Angele's class wondered if students who jump farther in the long jump usually jump higher in the high jump. They measured how long and how high each student jumped. Then they identified the median length for each type of jump.

The chart shows the students who were at or above the median and below the median for each type of jump.

Comparisons to Medians

Name	Akeem	Angele	Ayan	Chandra	Denise	Emilio	Isabella	James	Jorge	Khaled
Long jump	at or above	below	below	at or above	below	below	below	below	at or above	below
High jump	below	below	at or above	below	below	below	below	below	at or above	below

Name	Kurt	Li Ming	Maggie	Marc	Qi	Raven	Rebecca	Rodrigo	Tara	Tom
Long jump	at or above	at or above	at or above	below	below	at or above	below	at or above	at or above	at or above
High jump	at or above	below	at or above	below	at or above	at or above	at or above	at or above	at or above	at or above

? Do students who jump farther in the long jump also jump higher in the high jump?

Angele's Venn Diagram

I used a Venn diagram to sort the students. The red circle shows the long jumpers at or above the median and the blue circle shows the high jumpers at or above the median.

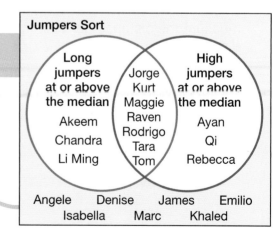

90

The area where the two circles overlap shows that 7 of the 20 students are at or above the median for each type of jump.

The area outside the two circles shows that 7 of the 20 students are below the median for each type of jump.

So my Venn diagram shows that students who jump farther in the long jump usually jump higher in the high jump. Also students who don't jump as far in the high jump usually don't jump as far in the long jump.

Khaled's Carroll Diagram

I used a **Carroll diagram** to sort the students because it is a good way to show relationships between two sets of data.

I am below the median for both types of jump. So my name is in the cell that shows below the median for each type of jump. Most jumpers are at or above median for both, or below median for both. My Carroll diagram shows that students who are good at the high jump are usually good at the long jump.

		Long Jump	
		At or Above	Below
High Jump	At or Above	Jorge, Kurt, Maggie, Raven, Rodrigo, Tara, Tom	Ayan, Qi, Rebecca
	Below	Akeem, Chandra, Li Ming	Angele, Denise, Emilio, Isabella, James, Khaled, Marc

Reflecting

1. What would Angele's Venn diagram look like if there was no relationship between the long jump and high jump? Explain your reasoning.

2. What would Khaled's Carroll diagram look like if students who were good long jumpers were always good high jumpers? Explain your reasoning.

3. What is the same about Angele's Venn diagram and Khaled's Carroll diagram? What is different?

Carroll diagram
A diagram that uses rows and columns to show relationships

Checking

4. The girls take 10 free throws with their stronger or dominant arm and 10 free throws with their opposite arm. The table shows the number of points scored by each student using each arm.
   a) Determine the mean and the median number of points scored by shooting with each arm.
   b) Use the two means or two medians to sort students into four groups in a Venn diagram or a Carroll diagram.
   c) Do students who score the most points with their dominant arm usually score the most points with their opposite arm? Use the diagram to justify your answer.

Practising

5. Copy and complete the Carroll diagram to sort the numbers from 1 to 20.

	Greater than 9	Less than or equal to 9
Even	10	
Odd		1

6. The table shows the rankings of 20 Canadian cities based on 30 years of weather records kept by Environment Canada.
   Whitehorse is the driest city, and Vancouver has the warmest winters.
   Use a Venn diagram or a Carroll diagram to determine whether the driest cities have the coldest winters. Show your work.

7. Gather other pairs of weather rankings from Environment Canada. Use a graphic organizer to determine if there is a relationship between the rankings.

Baskets Made

	Dominant	Opposite
Angele	5	4
Ayan	2	1
Chandra	3	3
Denise	7	5
Isabella	5	3
Li Ming	1	0
Maggie	2	1
Raven	4	1
Rebecca	0	0
Tara	6	5

Weather Ranking

City	Driest	Coldest Winter
Whitehorse	1	4
Yellowknife	2	1
Saskatoon	3	5
Kelowna	4	19
Regina	5	6
Calgary	6	10
Edmonton	7	8
Winnipeg	8	3
Thompson	9	2
Toronto	10	16
Ottawa	11	11
Montreal	12	12
Fredericton	13	9
Charlottetown	14	14
Vancouver	15	20
Moncton	16	13
Quebec City	17	7
Halifax	18	17
Sydney	19	15
St. John's	20	18

CHAPTER 3

Skills Bank

LESSON

1

1. A music Web site asked 20 customers how many hours they listened to music the previous week. The chart shows the results.
   a) Organize the data into a tally chart with intervals.
   b) Sketch a graph to display the data. Justify your choice of graph.
   c) Would the results of the survey represent the number of hours of music listened to by everyone? Explain your reasoning.

Music Survey Results (h)

3	10	15	6	8
12	4	4	7	10
2	1	1	21	10
16	17	14	18	10

2

2. a) Describe the location of each point plotted on the graph. Use coordinate pairs.
   b) Describe the location of two other points that are on the line. Use coordinate pairs.

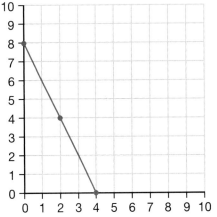

3

3. Emily bounces a basketball at a steady speed for 2 min. Alexis records the number of bounces after every 20 s. The table shows their results.
   a) Draw a line graph using the data in the table.
   b) Use the graph to predict the number of bounces after 50 s.
   c) Predict the number of bounces after 3 min. Extend the graph to check your prediction.

Basketball Bounces

Time (s)	Bounces
0	0
20	30
40	60
60	90
80	120
100	150
120	180

4

4. The scatter plot shows the number of attempted and completed passes by quarterbacks in the CFL. Does the scatter plot show that the more passes a quarterback attempts, the more completed passes he makes? Use the graph to explain your reasoning.

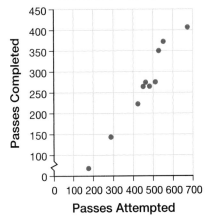

5. Determine the mean and the median of each set of data.
   a) 1, 0, 5, 5, 7, 2, 4, 2, 3, 3, 5, 7, 99
   b) 100, 99, 97, 98, 101, 96, 95, 98
   c) 2, 8, 6, 10, 10, 10, 10
   d) 250, 250, 246, 247, 251, 240, 242, 242

6. Austin's class measured the masses in kilograms of backpacks carried by 10 Grade 1 students, 10 Grade 3 students, and 10 Grade 6 students.
   Compare means or medians to determine which grade carries the heaviest backpacks.

   Backpack Masses (kg)
   Grade 1: 4, 5, 4, 3, 7, 5, 5, 6, 8, 3
   Grade 3: 7, 6, 8, 8, 7, 8, 3, 6, 9, 9
   Grade 6: 9, 10, 10, 5, 12, 10, 12, 11, 13, 8

7. A Web site asked people the question, "Do you listen to the commentaries on DVD movies?" The graph shows the results.
   a) Explain how the appearance of the graph would change if you doubled the size of each unit of the scale on the vertical axis.
   b) Explain how the appearance of the graph would change if you halved the size of each unit of the scale on the vertical axis.
   c) Explain how the appearance of the graph would change if you combined the first two categories and the last two categories.

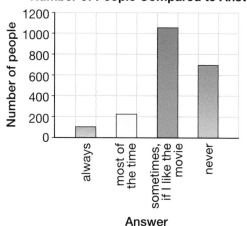

8. Hannah measured the number of heartbeats in 1 min of 12 girls when they were resting and when they were exercising.
   a) Determine the median resting heart rate and the median exercise heart rate.
   b) Use the two medians to create a Carroll diagram.
   c) Is there a relationship between a student's resting heart rate and exercise heart rate? Use your Carroll diagram to justify your reasoning.

Number of Heartbeats in 1 min

Student	Resting	Exercising
Hannah	66	151
Jasmine	120	155
Kaitlyn	80	157
Gabrielle	75	155
Hana	100	165
Anita	93	162
Natasha	85	140
Monique	110	169
Tessa	90	161
Jessica	130	177
Lauren	70	153
Ashley	60	149

CHAPTER 3
Problem Bank

LESSON

2

1. A red die and a blue die are tossed and the two numbers are used to create a coordinate pair. The point for each pair is plotted on this grid.
   a) Name a coordinate pair that will be inside the rectangle.
   b) Name a coordinate pair that will be outside the rectangle.
   c) What fraction of the coordinate pairs will be inside or on the sides of the rectangle? Explain your reasoning.
   d) What fraction of the coordinate pairs will be outside the rectangle? Explain your reasoning.

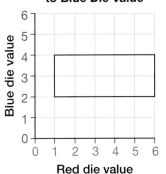

2. a) How would the red line segment look if the numbers in each coordinate pair were reversed?
   b) How would the green line segment look if the numbers in each coordinate pair were reversed?

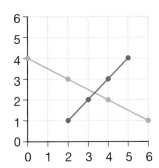

3

3. Carmen was thirsty when she came home from soccer practice. The line graph compares the amount of water in her glass and time. Describe what is happening in each part of her graph.

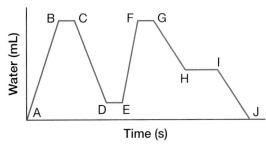

4. The scatter plot shows the heights and masses of five animals. Which point represents each animal? Explain.

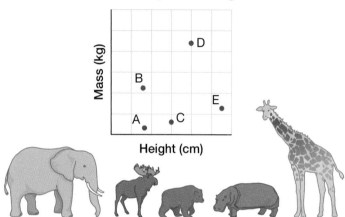

5. a) A biologist is studying the snowy owl population in Canada. She found nine nests containing three to nine eggs. The median number of eggs is six. Three nests contained five eggs. How many eggs might have been in each nest? Explain how you solved the problem.

   b) Explain how your answer would change if the mean number of eggs was also 6.

6. a) Calculate the mean of each set.
   2, 3, 5, 6
   4, 8, 10, 12, 1

   b) Add one number to one set so that the means of both sets are equal. Explain your reasoning.

7. Jordan made a Carroll diagram to organize the numbers from 1 to 20 but forgot to label each row and column.
   a) What labels should he write in each row and column?
   b) Draw a Venn diagram to organize Jordan's numbers.

	?	?
?	3, 6, 9, 12	1, 2, 4, 5, 7, 8, 10, 11
?	15, 18	13, 14, 16, 17, 19, 20

CHAPTER 3

Frequently Asked Questions

Q: How do you use the mean and median to compare sets of data?

A: You can use the mean or median to compare sets of data. For example, the table shows the number of minutes two brothers talked during each phone call in a weekend.

The mean or median can be used to compare the number of minutes each brother talks.
The mean is calculated by adding all times and then dividing by the number of calls.

The mean times seem to show that Rufus usually talks longer than Cedric.
The median can be determined by locating the middle number after the numbers are placed in order.

The median times seem to show that Cedric usually talks longer than Rufus.

Because the time of 197 min is much different from Rufus's other times, his mean time is much greater than Cedric's mean time. When the numbers are about the same in a set, you can use either the mean or median to compare sets. But if some numbers are quite different from the other numbers in a set, the median is sometimes a better choice.

Number of Minutes Talking		
Cedric		
20	30	20
25	15	30
23	22	13
Rufus		
1	5	1
4	2	197

	Total (min)	Number of calls
Cedric	198	9
Rufus	210	6

	Mean time
Cedric	198 ÷ 9 = 22
Rufus	210 ÷ 6 = 35

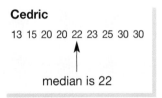

Cedric
13 15 20 20 22 23 25 30 30
median is 22

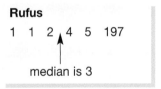

Rufus
1 1 2 4 5 197
median is 3

Q: How do Carroll diagrams organize data?

A: A Carroll diagram organizes data that can be split into opposites. For example, this Carroll diagram compares the scores of 10 students on two math tests. The Carroll diagram shows there is a relationship between scores on the first math test and scores on the second math test.

		Math Test 1	
		At or above the median	Below the median
Math Test 2	At or above the median	Denise, Akeem, Li Ming, Kurt	Tara
	Below the median	James	Marc, Chandra, Rodrigo, Angele

CHAPTER 3
Chapter Review

LESSON 1

1. A theatre owner asked 20 Grade 6 students buying a movie ticket how many movies they see each month. The chart shows the results.
   a) Organize the data into a tally chart with intervals.
   b) Sketch a graph to display the data. Justify your choice of graph.
   c) Would the results of the survey represent adults in the neighbourhood? Explain your reasoning.

Movie Survey Results

3	1	1	6	8
2	1	1	5	6
2	4	4	7	10
9	7	8	1	1

LESSON 2

2. a) Plot the coordinate pair (4, 5) on a grid.
   b) Draw a 2 unit by 2 unit square with the point (4, 5) in the middle. List all points that are on the perimeter of the square.

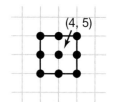

LESSON 3

3. A bug starts at a corner of this rectangle drawn on centimetre dot paper. It crawls clockwise around the rectangle until it ends up back at the start. The table shows the distance travelled from the start and the distance it had left to travel at each corner.

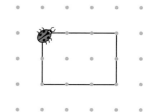

Crawling Bug Distance

Distance travelled (cm)	0	3	5	8	10
Distance to go (cm)	10	7	5	2	0

   a) Compare distance travelled to distance to go. Use a line graph.
   b) Describe the shape of the line graph.
   c) How far had the bug moved from the start when it had 3.5 cm left? Use the graph.

LESSON 4

4. Mika's Grade 6 class measured the leg length and height of each student. The table shows the data.

Class Measurements

Height (cm)	158	163	160	173	161	157	153	156	148	151	161	152	147	141	141	150	147	157	141	148
Leg length (cm)	77	79	78	78	76	76	75	75	75	74	75	72	72	71	71	71	71	75	68	70

   a) Compare leg length to height. Use a scatter plot.
   b) Do taller students usually have longer legs? Use the scatter plot to justify your reasoning.

5. Chelsea recorded the amount of sugar in a 30 g serving of each of 20 cereals. Ten were marked Kids' Cereal and ten were marked Family Cereal.
   a) Determine the median amount of sugar in each set.
   b) Which type of cereal typically has more sugar?

   Amount of Sugar in a 30-g Serving
   Kid's Cereal (g): 16, 12, 13, 12, 11, 14, 14, 9, 6, 13
   Family Cereal (g): 2, 3, 6, 16, 10, 2, 9, 5, 11, 6

6. Determine the mean and the median of each set of data.
   a) 3, 13, 15, 15, 17, 22, 24, 32, 33, 34, 89
   b) 7, 9, 5, 7, 9, 5, 9, 13
   c) 41, 106, 97, 348, 8
   d) 31, 77, 55, 44, 22, 77

7. The line graph shows the mean height of children at each age from 2 to 11 years.
   a) Why did the scale on the vertical axis start at 80 instead of 0?
   b) Predict the appearance of the line graph if the scale on the vertical axis changed from 1 unit for 10 cm to 1 unit for 5 cm.
   c) Explain how the appearance of the graph would change if the scale on the vertical axis changed from 1 unit for 10 cm to 1 unit for 20 cm.
   d) How can you use the graph to estimate the mean height of 12-year-olds?
   e) How can you use the graph to estimate the mean height of 8.5-year-olds?

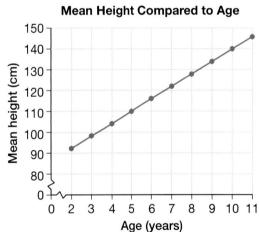

CHAPTER 3
Chapter Task

Investigating Body Relationships

Clothes makers study body measurements in order to make clothing that fits people of different sizes.

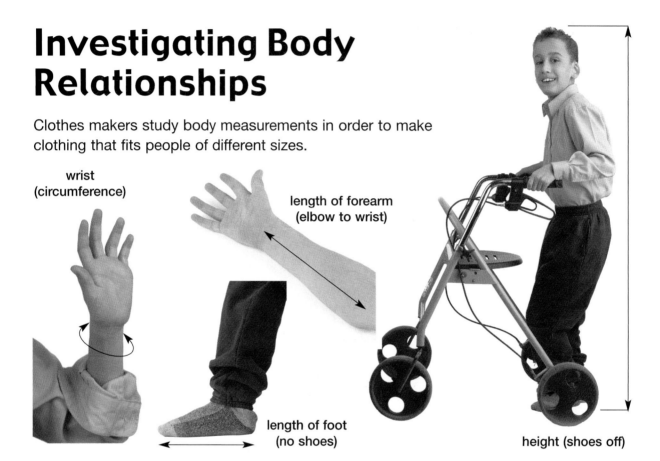

? **What is the relationship between two different body measurements?**

A. Choose a pair of measurements that you think might be related. Then use a measuring tape and rulers to measure all students in your class.

B. Record each pair of measurements of each student in a spreadsheet or graphing software. Determine the mean and median of each set of data.

C. Use your data to create a scatter plot, a Venn diagram, or a Carroll diagram.

D. Is there a relationship between the measurements in the pair you chose? Justify your answer.

Task Checklist
- ☑ Did you explain your thinking?
- ☑ Did you include enough detail?
- ☑ Did you use correct math language?
- ☑ Does your argument make sense?

Cumulative Review

Cross-Strand Multiple Choice

1. Which is the common difference?

   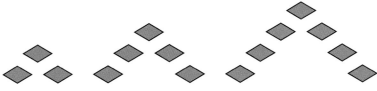

   1st arrangement   2nd arrangement   3rd arrangement

   A. 3   B. 2   C. 9   D. 5

2. Which is the sixth term in the pattern for pattern rule $7 \times n$?
   A. 49   B. 7   C. 42   D. 13

3. Which is the value of ■ so the two expressions in the equation $3 \times 3 = ■ - 1$ are equal?
   A. 10   B. 8   C. 12   D. 9

4. Which is the numeral represented by this counter model?

Millions			Thousands			Ones		
Hundreds	Tens	Ones	Hundreds	Tens	Ones	Hundreds	Tens	Ones
●● ●	●			●●●●●	●●			

   A. 301 042 000   C. 310 420 000
   B. 310 042   D. 310 042 000

5. Which numeral is five hundred fifty thousand sixty-two?
   A. 50 050 062   C. 505 620
   B. 550 062   D. 562 000

6. These numbers are ordered from least to greatest: 420 003, 420 499, 420 ■01, 420 503.
   Which is the missing digit?
   A. 5   B. 0   C. 4   D. 1

7. Which number is equal to 0.6 million?
   A. 6 000 000   C. 60 000
   B. 600 000   D. 60 000 000

8. Which is 5 hundredths + 7 thousandths in words?
   A. fifty-seven thousandths
   B. five and seven thousandths
   C. five hundred seven thousandths
   D. fifty-seven hundredths

9. Which is one thousandth greater than 3.209?
   A. 3.209   B. 3.309   C. 3.210   D. 3.219

10. Which is 0.806 rounded to the nearest hundredth?
    A. 0.80   B. 0.81   C. 0.90   D. 0.86

11. Which decimals are ordered from the least to the greatest?
    A. 1.003, 0.203, 0.230, 0.302
    B. 0.203, 0.302, 0.230, 1.003
    C. 0.230, 0.203, 0.302, 1.003
    D. 0.203, 0.230, 0.302, 1.003

12. Which are the coordinates for the point on the coordinate grid?
    A. (0, 4)   B. (1, 3)   C. (1, 4)   D. (1, 5)

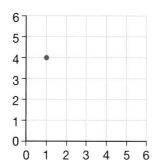

13. Angele made this graph to show the distance she travelled in a bus along a highway from North Bay. Which is the distance Angele had travelled after 3 h 30 min?
    A. 125 km   B. 225 km
    C. 150 km   D. 175 km

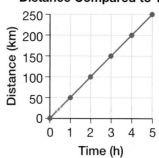

Distance Compared to Time

14. The mean of these numbers is 340:
    ■, 340, 380, 280, 340. Which is ■?
    A. 360   B. 335   C. 340   D. 272

Cross-Strand Investigation

Lizards and Snakes

Ayan, Qi, and Isabella are researching data about lizards and snakes for their projects about reptiles.

Komodo dragon

15. a) Qi read that the largest lizard is the Komodo dragon whose mass is up to 0.135 t. Write the mass in expanded form.
    b) Komodo dragons can be 2.860 m long. Round the length to the nearest tenth.
    c) The smallest lizard is a gecko called the Jaragua sphaero. It's 0.016 m long. Write the length in words.
    d) Ayan found information about lizards. The fastest lizard, the spiny-tailed iguana, can travel 9 m each second. Use a variable to write an expression to show the distance for any number of seconds.
    e) Make a table to show the distance a spiny-tailed iguana can travel for 0 s to 10 s.
    f) Use the data in your table to create a line graph. Graph the distance compared to the number of seconds.
    g) Use your graph to estimate the distance a spiny-tailed iguana can travel in 5.5 s.
    h) What would happen to your graph if you increased the value of each unit on the scale on the vertical axis?

spiny-tailed iguana

16. a) Isabella read about the snake with the longest fangs, a gaboon viper. It can be 1.830 m long, and have fangs 0.051 m long. Write each number in expanded form.
    b) She learned that the fastest snake is a black mamba. It can travel 5.3 m in a second. Write a pattern rule for calculating a distance it can travel in any length of time. Justify your pattern rule.
    c) How far can a black mamba travel in 15 s?
    d) Create a survey question about snakes or other reptiles.
    e) Suppose you used your question for a survey. Do you think the results would be the same for your classmates as for adults in your community? Explain.

gaboon viper

black mamba

Moths and Other Insects

Tara and Jorge choose insects for their projects.

17. **a)** Jorge researched this data about moth wingspans. What is the mean for the greatest wingspans, to the nearest millimetre?

Moth Wingspan Ranges

Moth	Least (mm)	Greatest (mm)	Moth	Least (mm)	Greatest (mm)
regal	120	150	sheep	60	70
rosy maple	30	50	cecropia	120	150
rattlebox	33	46	artichoke plume	20	28
lichen	25	30	cynthia	75	135
luna	80	115	hummingbird	38	50
imperial	100	150	white-lined	65	90
sod webworm	12	39	tomato hornworm	90	110
big poplar sphinx	90	140	wild-cherry	75	115
measuring worm	8	65	California cankerworm	25	35
giant silkworm	30	150	western tent caterpillar	32	41

b) Use a scatter plot to compare greatest wingspans to least wingspans.

c) Can you use your scatter plot to predict the greatest wingspan of a type of moth if you know its least wingspan? Use your scatter plot to justify your reasoning.

d) Tara read that large insects appeared about 345 000 000 years ago, and modern insects appeared about 280 000 000 years ago. Write the expanded form and the words for each number.

e) The insect with the slowest wingbeat is the swallowtail butterfly. Its wingbeat is 300 each minute. Use a table to show the number of wingbeats from 0 min to 5 min.

f) How many times do the wings of a swallowtail butterfly beat in half an hour? Use a pattern rule.

g) The most massive insects are the goliath beetle with a mass up to 0.100 kg and a weta with a mass up to 0.071 kg. What is the most massive insect?

h) The longest insect is the giant walking stick. The longest specimen on record was 0.555 m. Write the expanded form and the words for this length.

luna moth

giant walking stick

CHAPTER 4

Addition and Subtraction

Goals

You will be able to

- use mental math and pencil and paper to add and subtract whole and decimal numbers
- estimate sums and differences of whole and decimal numbers
- use addition and subtraction to solve problems
- justify your choice of calculation method

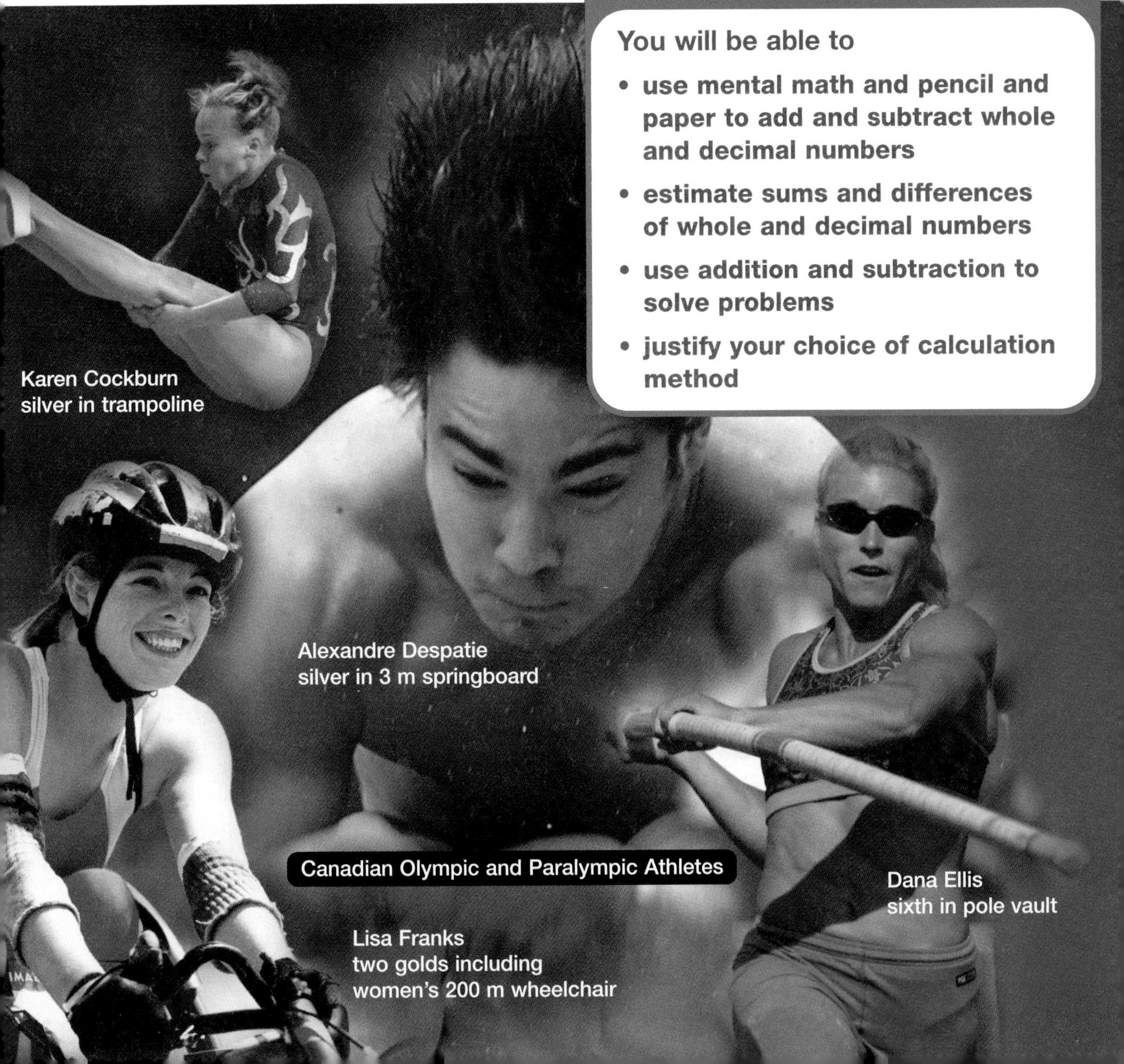

Karen Cockburn
silver in trampoline

Alexandre Despatie
silver in 3 m springboard

Canadian Olympic and Paralympic Athletes

Lisa Franks
two golds including
women's 200 m wheelchair

Dana Ellis
sixth in pole vault

CHAPTER 4

Getting Started

Planning Cross-Country Routes

A cross-country competition is being held along the paths of the Meewasin Trail in Saskatoon. Junior student races are about 5 km long. Senior student races are about 10 km long.

? **Which routes can you create for Junior and Senior events?**

A. One senior race route is along the west path from the first bridge to the sixth bridge and back. Estimate how long the route is. Explain how you estimated.

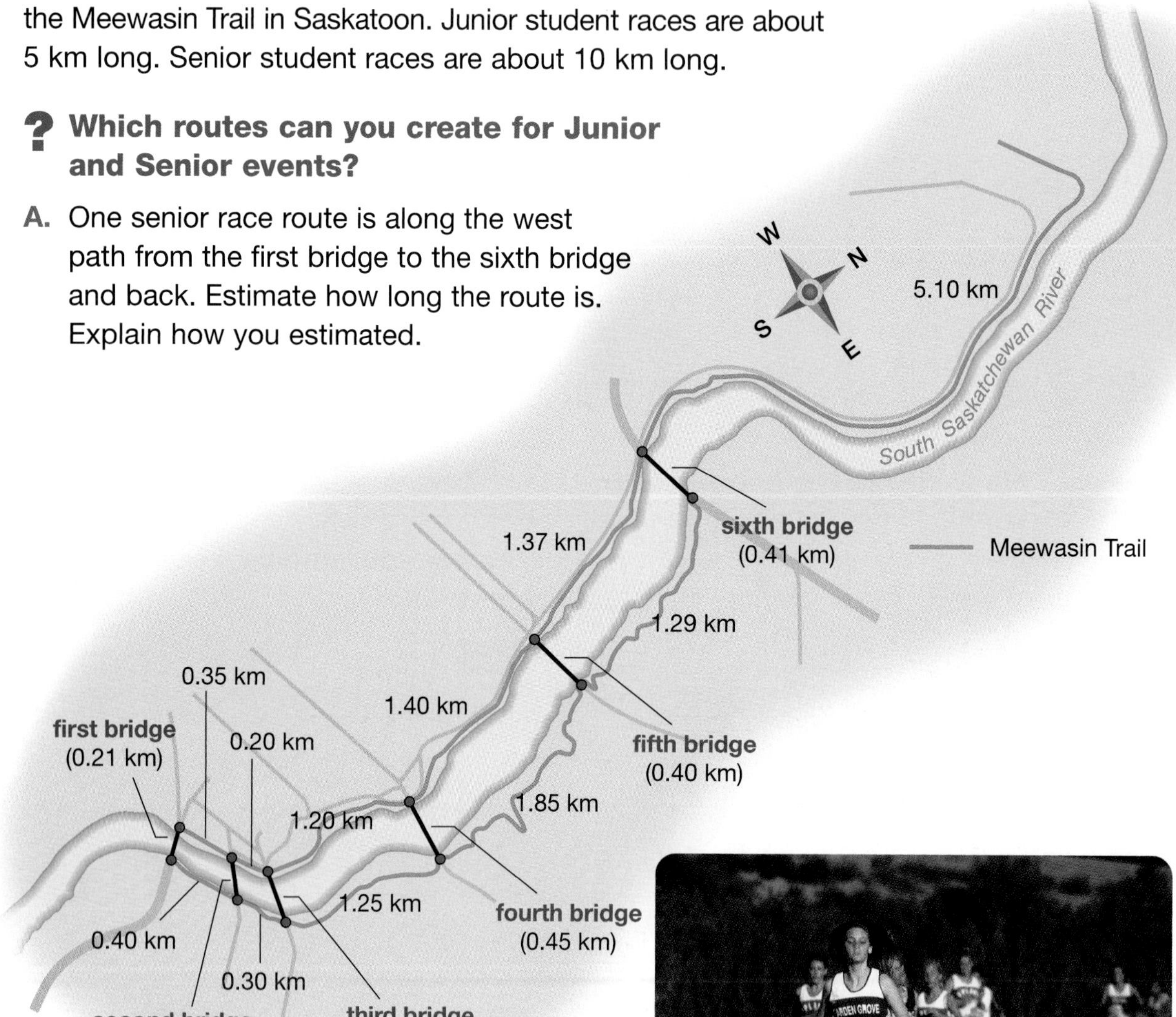

Map labels:
- 5.10 km
- South Saskatchewan River
- sixth bridge (0.41 km)
- 1.37 km
- 1.29 km
- Meewasin Trail
- 0.35 km
- 1.40 km
- fifth bridge (0.40 km)
- first bridge (0.21 km)
- 0.20 km
- 1.20 km
- 1.85 km
- 0.40 km
- 1.25 km
- fourth bridge (0.45 km)
- 0.30 km
- second bridge (0.17 km)
- third bridge (0.41 km)

106

B. If a junior student ran on the west path from the sixth bridge to the fourth bridge, how much farther would she have to run to go 5 km? Show your work.

C. Did you use mental math, pencil and paper, or a calculator to answer Part B? Justify your choice of method.

D. One junior race route is along the west path from the first bridge to the sixth bridge. Another junior race route starts at the west side of the sixth bridge, crosses the bridge, and follows the east path to the first bridge. Which route is longer and by how much?

E. Create and describe a new race route for junior students and a new race route for senior students. Show your work.

F. How much longer is your route for the senior students than your route for the junior students? Show your work.

Do You Remember?

1. Use mental math.
   a) $25 + 76$
   b) $99 + 199$
   c) $1000 - 895$
   d) $233 - 99$

2. Use estimation to determine if each calculation is reasonable. Correct any unreasonable calculations.
   a) $2365 + 4869 = 7234$
   b) $5497 + 3476 = 7973$
   c) $6238 - 1639 = 5599$
   d) $8000 - 1567 = 6433$

3. Calculate.
   a) $4394 + 2186 + 6534 = \blacksquare$
   b) $7.5 + 1.7 = \blacksquare$
   c) $2000 - 757 = \blacksquare$
   d) $2.65 - 0.88 = \blacksquare$

4. Hailey won a $100 gift certificate for bowling.
   a) About how many friends can Hailey invite to a bowling party? Explain.
   b) Calculate the total cost of the bowling party.
   c) How much money will Hailey have left on her gift certificate?

How about a BOWLING PARTY?

Great fun • Great value
Only $11.95 per child (tax included)

CHAPTER 4

1 Adding and Subtracting Whole Numbers

Goal Use mental math strategies to calculate sums and differences.

Qi plays the game "Odd and Even" with two coins. The outcomes 'both heads' and 'both tails' are called "even." The outcome 'one head, one tail' is called "odd." He wants to estimate the **frequency** of each of the three outcomes, so he uses an online coin tosser to toss two coins 1000 times. The bar graph shows the results of his experiment.

? How many times might each outcome occur?

A. In Qi's experiment, the frequency of 'one head, one tail' was 510 in 1000 tosses. Use mental math to determine the possible frequencies of 'both heads' and 'both tails'. Describe your strategy.

B. In another experiment, the frequency of 'both tails' was 270 in 1000 tosses. What might be the frequencies of 'both heads' and 'one head, one tail'? Use mental math. Describe your strategy.

C. If you tossed a coin 2000 times, what might the frequency of each outcome be? Use mental math. Describe your strategy.

D. Choose another number of tosses greater than 1000. What might the frequency of each outcome be? Use mental math. Describe your strategy.

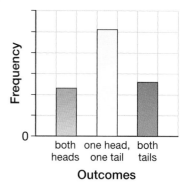

frequency
The number of times an event occurs

Reflecting

1. Did you use the same strategies for Parts A and B? Explain.

2. Did the number of tosses you chose for Part D allow you to use mental math? Explain why or why not.

Curious Math

Number Reversal

Rodrigo says that you will always get the number 6801 if you follow his instructions.

Rodrigo's Instructions

Step 1 Write any three-digit number. It should contain different digits in the hundreds and ones places.

Step 2 Reverse the digits. Write the new number.

Step 3 Subtract the lesser number from the greater number. If the result is less than 100, put zeros in front to make a three-digit number. For example, if the difference is 27, use 027.

Step 4 Reverse the digits from Step 3. Write the new number.

Step 5 Add the numbers from Step 3 and Step 4.

Step 6 Turn your total upside down.

1. Follow Rodrigo's instructions. What number did you get?

2. Choose three other numbers and follow Rodrigo's instructions. What number did you get each time?

3. For Step 1, choose a decimal number in the form ■.■■ instead. What number did you get?

CHAPTER 4

2 Estimating Sums and Differences

You will need
- number lines

Goal: Estimate sums and differences to solve problems.

For her science presentation, Tara described the stages of a flight of SpaceShipOne in 2004. The rocket was in a contest for a $10 million prize. To win, it had to reach a height of 100 000 m in two flights in a two-week period.

? Did the rocket reach a height greater than 100 000 m?

Stages in the September 29 Flight

Stage 1	The White Knight jet carried the rocket up to a height of 14 173 m.
Stage 2	The White Knight released the rocket. The rocket's engine lifted the rocket another 40 691 m higher.
Stage 3	The rocket coasted another 47 853 m higher. Then it glided safely back to Earth.

Tara's Solution

I'll use a number line to estimate how much higher the rocket must go to reach 100 000 m.

The height of the rocket in Stage 1 was 14 173 m. I will round 14 173 m to 14 000 m and mark this height on the number line.

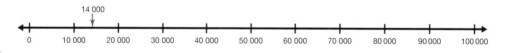

A. Copy Tara's number line. Use the number line to estimate how much higher the rocket has to go to reach 100 000 m.

B. Use the number line to estimate the height of the rocket at the end of Stage 2.

C. Use the number line to estimate how much higher the rocket has to go to reach 100 000 m.

D. Did the rocket reach a height greater than 100 000 m during Stage 3? Explain your reasoning.

Reflecting

1. Why were you able to solve this problem using only estimation?
2. Did you round to the nearest tens, hundreds, or thousands to solve this problem? Explain.

Checking

3. Tara also described the stages of the SpaceShipOne flight on October 4.
   a) Estimate the height of the rocket at the end of Stage 2. Show your work.
   b) At the end of Stage 2, about how much higher does the rocket need to go to reach 100 000 m?
   c) Did the rocket reach 100 000 m and win the $10 million prize? Explain your reasoning.

Stages in the October 4 Flight

Stage 1	The jet carried the rocket to a height of 14 356 m.
Stage 2	The rocket's engine lifted the rocket up another 50 566 m.
Stage 3	The rocket coasted up another 47 092 m. Then it glided safely back to earth.

Practising

4. The Marianas Trench in the Pacific Ocean is the deepest spot in any ocean. It has a depth of 10 924 m. The Japanese research submarine *Shinkai 6500* has reached a depth of 6526 m. About how much farther would it need to dive to reach the bottom of the Marianas Trench? Use estimation to solve the problem. Describe your strategy.

5. The chart shows the number of people in different age groups living in Mississauga, Ontario.
   a) About how many people are 24 years old or younger?
   b) About how many more people are in the over 25 years old group than are in the 24 years old or younger group?
   c) About how many people live in Mississauga?

Age	Number
0–4	40 020
5–14	90 445
15–19	43 860
20–24	41 595
25+	397 005

CHAPTER 4

3 Adding Whole Numbers

You will need
- a calculator

Goal Solve problems by adding four 3-digit whole numbers.

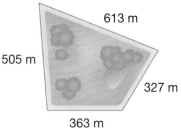

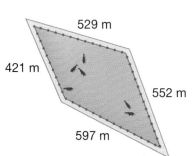

Ayan runs around a park three times each day.
Isabella lives in the country and runs around a horse paddock two times each day.

? Who jogs farther each month and by how much?

Ayan's Addition

The perimeter around the park is about 1800 m.

I can calculate the actual perimeter by adding. I will start with the ones column.

```
      1
    6 1 3
    3 2 7
    3 6 3
  + 5 0 5
  ---------
          8
```

Isabella's Addition

The perimeter of the paddock is about 2100 m.

I can calculate the actual perimeter by adding. I will start by adding the numbers in the hundreds column.

```
    5 2 9
    5 5 2
    5 9 7
  + 4 2 1
  ---------
  1 9 0 0
    1 8 0
```

A. Calculate the perimeter of the park and the paddock using each girl's method. Are your answers reasonable? Explain.

B. Estimate and then calculate the distances that Ayan and Isabella run each day. Show your work.

C. Estimate and then calculate the distances Ayan and Isabella run in a 30-day month.

D. Who runs farther in a 30-day month?

E. About how much farther does one girl run than the other in a 30-day month?

Reflecting

1. Explain how Ayan and Isabella might have estimated the perimeters.

2. a) Why did Ayan record the digit 1 in the tens column?
   b) Why did Isabella record 1900 and 180 in her addition?

3. Which addition method did you prefer for this problem? Explain.

Checking

4. In the summer, Kurt runs once every day around the route shown. In the winter, he runs twice around the same route three times a week.

   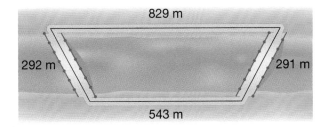

   a) Estimate the distance he runs each week in the summer and winter. Show your work.
   b) Calculate the distance he runs each week in the summer and winter. Do your answers seem reasonable? Explain.
   c) Does he run a greater distance each week in the winter or in the summer? About how much greater is the distance?

Practising

5. A neighbourhood has five elementary schools with a total enrollment of 1367 students. Four schools have enrollments of 387, 175, 245, and 116.
   a) What is the enrollment of the fifth school?
   b) Use estimation to check if your answer is reasonable.

6. To raise money, Dylan's school put in five vending machines selling fruit juices. The machines accept only quarters. At the end of one month, four of the vending machines had 400, 280, 360, and 240 quarters in them. The total money in all five machines was $400.
   a) Estimate and then calculate the number of quarters in the fifth machine. Show your steps.
   b) Did you use mental math, pencil and paper, or a calculator to calculate each step? Justify your choices.

CHAPTER 4

4 Subtracting Whole Numbers

You will need
- counters (3 different colours)
- a decimal place value chart
- a calculator

Goal Subtract whole numbers to solve problems.

Jorge is making a poster for Heritage Day at his school. He made a bar graph comparing the number of people in Ontario who spoke Ojibway in 1996 and in 2001.

? How many more people spoke Ojibway in 1996 than in 2001?

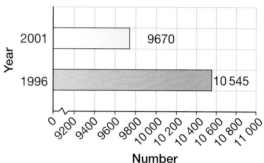

Ojibway Speakers in Ontario

Jorge's Subtraction

I estimate about 900 more people spoke Ojibway in 1996 than in 2001.

To calculate an exact answer, I'll subtract 9670 from 10 545.

Step 1 I use counters to model 10 545.

```
  10 545
-  9 670
```

Thousands			Ones		
Hundreds	Tens	Ones	Hundreds	Tens	Ones

Step 2 I regroup 10 000 as 9000 + 10 hundreds.
Then I add 1000 to 500 to get 1500.

```
   9 15
  10̶ 5̶45
-  9 670
```

Thousands			Ones		
Hundreds	Tens	Ones	Hundreds	Tens	Ones

Maggie's Subtraction

I can use mental math to solve the problem.
I first add 330 to each number.

10 545 + 330
− 9670 + 330
──────────

A. Complete Jorge's and Maggie's subtractions.
B. Compare both answers to Jorge's estimate. Are your answers reasonable?
C. How many more people spoke Ojibway in Ontario in 1996 than in 2001?

Reflecting

1. Why did Jorge regroup 10 000 as 9000 + 10 hundreds?
2. Instead of estimating, Maggie says she can use addition to check her answer. Do you agree? Explain.
3. a) Why did Maggie choose to add 330 to each number in her subtraction?
   b) Would Maggie's answer have been the same if she added a number other than 330 to each number? Use an example to help you explain.

Checking

4. Sakima researched the number of people in Ontario who speak an Aboriginal language and how many of those speak Cree.
   a) In 1996, how many people in Ontario spoke an Aboriginal language other than Cree?
   b) In 2001, how many people in Ontario spoke an Aboriginal language other than Cree?
   c) Did you use mental math, pencil and paper, or a calculator to calculate? Justify your choice.
   d) Use addition or estimation to determine if your answers are reasonable. Show your work.

	1996	2001
Total aboriginal language speakers	21 420	19 970
Cree speakers	5445	4385

Practising

5. Calculate. Determine if your answers are reasonable. Show your work.
   a) 10 000 − 4555 = ■
   b) 6500 − 4992 = ■
   c) 78 320 − 9994 = ■
   d) ■ − 8045 = 80 000

6. There were 10 651 athletes in the Sydney 2000 Olympics. Of the athletes, 4069 were female. Estimate and then calculate the number of athletes who were male. Show your work.

7. In the 2004 Paralympics, there were 4000 athletes. In the 2000 Paralympics, there were 3843 athletes. Estimate and calculate how many more athletes were in the 2004 Paralympics than in the 2000 Paralympics. Show your work.

8. In the 2004 Olympics, there were 11 099 athletes. In the 2000 Olympics, there were 10 651 athletes. How many more athletes were in the 2004 Olympics than in the 2000 Olympics? Estimate to check your answer. Show your work.

9. Create a problem involving the subtraction of two multi-digit numbers. Solve your problem.

10. Each letter represents only one digit. What digit does each letter represent? Show your work.

    ```
      MONEY
    − SEND
      MORE
    ```

11. a) Write down a four-digit whole number with four different digits. Rearrange the digits to make the greatest and the least four-digit whole numbers.
    b) Subtract the least number from the greatest number. Estimate to check your answer.
    c) Use your answer to part b) to make the greatest and least numbers. Repeat these steps until you start to get the same number each time.
    d) Compare your final number in part c) to the final numbers of other students. Describe what you notice.

Curious Math

Subtracting a Different Way

Before he came to Canada, Khaled learned this method to subtract two whole numbers.

Step 1 To subtract 567 from 1234, I first add 10 to the top number and 10 to the bottom number.

TH	H	T	O	
1	2	3	4	+ 10 ones
-	5	6	7	+ 1 ten

I add 10 ones to the 4 ones and 1 ten to the 6 tens.

Now I can subtract 7 ones from from 14 ones.

TH	H	T	O
1	2	3	¹4
-	5	⁷6̶	7
			7

Step 2 Next I add 100 to the top number and 100 to the bottom number.

TH	H	T	O	
1	2	3	¹4	+ 10 tens
-	5	⁷6̶	7	+ 1 hundred
			7	

I add 10 tens to the 3 tens and 1 hundred to the 5 hundreds.

Now I can subtract 7 tens from 13 tens.

TH	H	T	O
1	2	¹3	¹4
-	⁶5̶	⁷6̶	7
		6	7

1 Complete Khaled's method.

2 Use Khaled's method to calculate each difference. Show your work.
a) 4526 − 778   c) 6205 − 423
b) 3000 − 356   d) 5600 − 1453

CHAPTER 4

Frequently Asked Questions

Q: How can you check whether an addition or subtraction calculation is reasonable?

A: Round the numbers and then use mental math to add or subtract the rounded numbers.

Suppose you used a calculator to subtract 152 799 from 248 365 and got this answer:

`95566.`

To check the answer, round each number to the nearest 50 000. Then use mental math to subtract:
250 000 − 150 000 = 100 000

The calculator answer of 95 566 is close to the estimate of 100 000, so the answer is reasonable.

Q: How can you add four 3-digit numbers?

A. There are many ways. Two are shown here. You can add the ones, tens, and hundreds digits from the right and regroup as you go or you can add the hundreds, tens, and ones from the left.

Add from the right and regroup.	Add from the left.
11 605 481 235 + 175 ───── 1496	605 481 235 + 175 ───── 1300 180 + 16 ───── 1496

Q: How do you subtract a 4-digit number from a 5-digit number?

A. You can subtract in different ways depending on the numbers.

Subtract by regrouping.	Add on to each number.

12 13
   11 13 14 15
  1̶ 2̶ 3̶ 4̶ 5̶
−   5 9 9 9
─────────
    6 3 4 6
	12 345 + 1 = 12 346
	− 5 999 + 1 = − 6 000
	──────────────────
	               6 346

Mid-Chapter Review

LESSON 1

1. Use mental math to calculate each answer. Describe your strategy.
   a) 650 + 350
   b) 175 + 225
   c) 99 + 59
   d) 1000 + 2500

2. Explain how knowing that 3000 + 4000 = 7000 can help you solve each problem.
   a) The attendance at a concert was 2999 the first night. The second night it was 3999. What was the total attendance of the two nights?
   b) One home theatre system costs $7000. Another one costs $2999. What is the difference in the two prices?
   c) What is the height of a plane flying 3005 m higher than 3999 m?

LESSON 2

3. Use estimation to decide which answers are reasonable. Show your work. Correct the unreasonable answers.
   a) 234 + 428 + 189 + 245 = 1096
   b) 999 + 999 + 999 + 999 = 2996
   c) 15 236 + 25 018 = 40 254
   d) 64 978 + 39 107 = 114 085

LESSON 3

4. Calculate. Show your work.
   a) 456 + 287 + 219 + 549
   b) 145 + 901 + 791 + 319
   c) 333 + 444 + 555 + 666
   d) 119 + 245 + 789 + 802

5. A four-day ball tournament is held annually. The table shows the attendance in 2004 and 2005. Which year had the greater attendance?

	Day 1	Day 2	Day 3	Day 4
2004	345	467	709	987
2005	456	139	815	932

LESSON 4

6. Taylor's goal is to walk 10 000 steps each day. One day, she walked 3456 steps and stopped to pet a dog. Then she walked 5365 steps and stopped to tie her shoes. How many steps does she have left to walk to reach her goal?

7. An online discussion forum for kids has a goal of 20 000 postings. In January, it had 1535 postings. In February, it had 2865 postings. In March, it had 3451 postings. How many more postings does it need to get to reach its goal?

8. Create and solve a subtraction problem using the two whole numbers 10 000 and 3998.

5 Adding and Subtracting Decimal Numbers

You will need
- play money

Goal Use mental math strategies to calculate sums and differences.

Rebecca's school is having a yard sale to raise funds for Oxfam Canada. Students donated the items for sale.

? How much will the items cost and how much change will you receive?

- inline skates $6.95
- hockey stick $3.75
- mixed shells $0.25
- robot blocks $1.95
- bracelet $1.25
- watch $5.99
- 12 magazines $2.50
- novel $0.50
- two-way radios $7.25
- movie $1.85

A. Rebecca has a $10 bill to spend. Select three items she can buy. Use mental math to calculate the total cost and her change. Describe your strategy for each calculation to another student.

B. Tom has a $10 bill and a $5 bill to spend. Select three items he can buy. Decide what bills he could use to make his purchases. Use mental math to calculate the total cost and his change. Describe your strategy for each calculation to another student.

C. Write the names and prices of three items you might donate to a yard sale. Make each price between $1.00 and $20.00.

D. Ask a classmate to calculate the total cost of your three items using mental math and describe his or her strategy.

E. Ask a classmate to provide enough play money to purchase your items. Make change using mental math. Describe your strategy to your partner.

Reflecting

1. Compare your mental math strategy to calculate the total cost of the three items in Part B to the strategies of your classmates. Describe one strategy that is different from yours.

2. Did the prices of the items you listed for Part C allow you and your partner to use mental math to calculate the total cost and the change? Explain why or why not.

Math Game

Mental Math with Money

You will need
- a money spinner
- a digit spinner
- a pencil
- a paper clip
- play money
- a calculator

Number of players: 2 to 4

How to play: Use mental math to calculate the change after buying an item. If you don't have enough money, use mental math to calculate how much more you need.

Step 1 Each player spins the money spinner once to find out which bills he or she has to spend.

Step 2 One player spins the digit spinner four times to make the price of an item. The first spin is the tens digit, the second spin is the ones digit, the third spin is the tenths digit, and the fourth spin is the hundredths digit.
$■■.■■

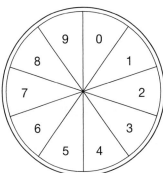

Step 3 If a player has enough money to buy the item, the player uses mental math to calculate the change. If a player does not have enough money, the player uses mental math to calculate how much more money he or she needs.

Step 4 Each player uses a calculator to verify his or her mental math calculation.

Step 5 If a player's mental math calculation is correct, he or she scores one point.

The first player with five points wins.

Emilio's Play

I spun a $20 bill and a $5 bill, so I have $25 to spend.

The price of the item was $78.29.

I used mental math to calculate how much more money I would need.

$78.29 − $25 is the same as $75.29 + $3 − $25.

$75.29 − $25 + $3 = $50.29 + $3 = $53.29

My answer is correct, so I get one point.

CHAPTER 4

6 Adding Decimals

You will need
- base ten blocks
- a decimal place value chart
- a calculator

Goal Add decimals using base ten blocks and pencil and paper.

For an Earth Day project, James measured the mass of newspapers and flyers that his family received in the mailbox each week.

Week	1	2	3	4
Newspapers (kg)	1.469	1.070	1.268	1.452
Flyers (kg)	1.610	0.978	1.164	1.012

? About how many kilograms of newspapers and flyers will James's family receive in one year?

James's Addition

I have the data for four weeks, so I will add to calculate the mass for one month. Then I will multiply the mass for one month by 12 to estimate the mass for one year.

I'll start by calculating the total mass for Week 1. Both masses are about 1.5 kg, so I estimate that the total mass is close to 3 kg.

I'll use blocks to represent these amounts.

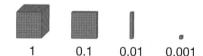

1   0.1   0.01   0.001

Step 1   I model 1.469 kg and 1.610 kg.

    1.469
  + 1.610

Step 2   I add the thousandths and the hundredths. I don't need to regroup.

    1.469
  + 1.610
       79

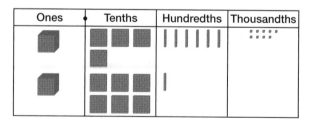

A. Complete James's addition to calculate the mass of paper for Week 1. Show your work.

B. Calculate the mass of paper received for each other week. Use estimation to check your answers. Show your work.

C. Calculate the mass of paper received in one month. Use estimation to check your answer. Show your work.

D. About how many kilograms of paper will James's family receive in one year? Show your work.

Reflecting

1. Could James's problem be modelled using a flat instead of a cube to represent one? Explain why or why not.
2. Why didn't James need to regroup when he added the thousandths and hundredths?
3. How is adding decimals like adding whole numbers? How is it different?

Checking

4. Chandra measured the mass of the newspapers and flyers her family received for one month.

Week	1	2	3	4
Newspapers (kg)	2.535	1.672	2.442	2.618
Flyers (kg)	0.599	0.476	0.603	0.253

   a) In one month, did James's family or Chandra's family receive a greater mass of newspapers and flyers? Show your work.
   b) Use estimation to check if your calculations are reasonable.

Practising

5. Calculate. Determine if your answers are reasonable. Show your work.
   a) 4.55 + 0.77   b) 1.5 + 4.678   c) 0.965 + 0.378   d) 2.769 + 1.569

6. Marcus used a Web site to measure his reaction time in thousandths of a second. Calculate the total reaction times for each hand. Which hand was faster?

   Right hand (seconds)   0.371   0.333   0.349   0.392   0.344
   Left hand (seconds)   0.375   0.359   0.422   0.429   0.336

7. Create a problem that involves the addition of two or more decimal numbers in thousandths. Solve your problem.

CHAPTER 4

7 Subtracting Decimals

You will need
- base ten blocks
- a decimal place value chart

Goal Subtract decimals using base ten blocks and pencil and paper.

Angele is preparing a report on gain in mass of babies in their first year. She used her own baby records to make a broken-line graph.

? In which three-month period did Angele have the greatest change in mass?

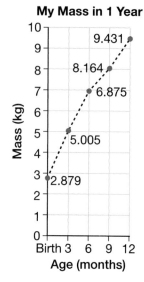
My Mass in 1 Year

Angele's Solution

To calculate my gain in mass for the first three-month period, I need to subtract my birth mass from my mass at three months.

I estimate that the gain was between 2 kg and 3 kg.

I'll use blocks to represent these amounts.

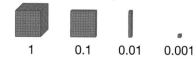
1   0.1   0.01   0.001

Step 1 I model 5.005 kg, my mass at three months.

5.005
− 2.879

Ones	Tenths	Hundredths	Thousandths
(5 ones blocks)			(5 thousandths)

Step 2 When I finished regrouping, my blocks looked like this.

4 9 9 15
5̶.̶0̶0̶5̶
− 2.879

Ones	Tenths	Hundredths	Thousandths
(4 ones)	(9 tenths)	(9 hundredths)	(15 thousandths)

124

Marc's Solution

I can use mental math to calculate the gain in mass.
I add 0.121 to each number.

$$\begin{array}{r} 5.005 \ + 0.121 \\ -\ 2.879 \ + 0.121 \\ \hline \end{array}$$

A. Complete Angele's and Marc's solutions. How much mass did Angele gain in her first three months?

B. Calculate the gain in mass in each of the other three-month periods. Show your work.

C. Use addition to check your answers in Part B. Show your work.

D. Which three-month period showed the greatest change in mass?

Reflecting

1. a) Why did Angele regroup 5.005 as 4 ones, 9 tenths, 9 hundredths and 15 thousandths?
   b) Explain the meaning of each number she wrote in each place value column above the number 5.005.

2. Explain why Marc added 0.121 to each number.

3. How is subtracting decimals like subtracting whole numbers? How is it different?

Checking

4. Angele used her brother's baby records to make another graph.
   a) In which three-month period did Angele's brother have his greatest gain in mass? Show your work.
   b) Use estimation to check if your calculations are reasonable. Describe your estimation strategy to another student.

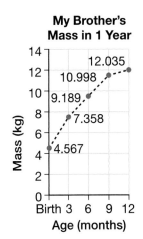

My Brother's Mass in 1 Year

Practising

5. Calculate. Determine if your answers are reasonable. Show your work.
   a) 4.0 − 1.4
   b) 6.05 − 2.38
   c) 3.000 − 0.537
   d) 6.050 − 0.9

6. Chloe is reading the park map at post 1. She wants to go fishing at post 4. Estimate and then calculate the difference in distance between the yellow and blue routes. Show your work.

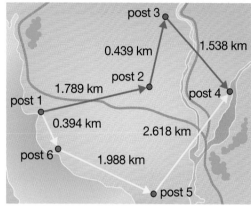

7. In the 2004 Olympics in Athens, Greece, Lori-Ann Muentzer won a gold medal in sprint cycling. She finished her first sprint in 12.126 s and her second sprint in 12.140 s. Calculate the difference in time between her two sprints. Show your work.

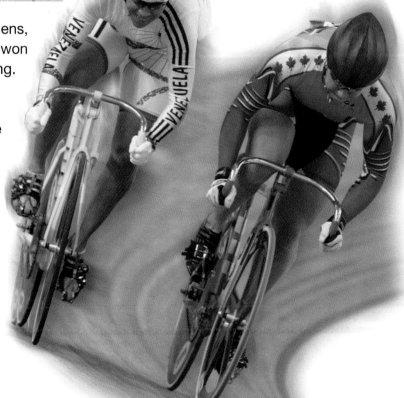

8. Create a problem that involves the subtraction of two decimal numbers. Solve your problem.

Mental Math

Using Whole Numbers to Add and Subtract Decimals

You can add or subtract decimals by thinking of the nearest whole number.

To add 9.8 to 5.7, add 10 to 5.7 and then subtract 0.2.

5.7 + 9.8 = ■

5.7 + 10 = 15.7
15.7 − 0.2 = 15.5

To subtract 4.98 from 15, subtract 5 from 15 and then add 0.02.

15 − 4.98 = ■

15 − 5 = 10
10 + 0.02 = 10.02

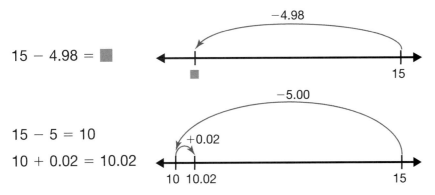

A. What are some other ways to add 9.8 to 5.7?

B. What are some other ways to subtract 4.98 from 15?

Try These

1. Calculate. Show your work.
   a) 8.8 + 9.8
   b) 14.9 + 12.3
   c) 2.99 + 12.58
   d) 6.0 − 1.9
   e) 10.5 − 4.8
   f) 13.45 − 6.98

CHAPTER 4

8 Communicate About Solving a Multi-Step Problem

You will need
- a calculator

Goal Explain a solution to a problem.

Li Ming has three flags to hang on a 5.75 m wire. She wants the end flags to be 0.50 m from the end of the wire. She also wants the flags to be an equal distance from each other. What will be the distance between each flag?

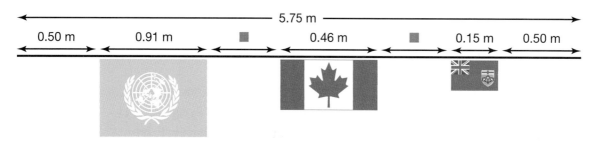

? How can Li Ming explain her solution to the problem?

Li Ming's Written Solution

I'll use the problem-solving steps Understand, Make a Plan, and Carry out the Plan to describe my solution.

Understand the Problem
I need to determine how much space there is between each flag.

Make a Plan
I'll add the two distances from the end of each flag to the end of each wire. I will subtract that sum from the length of the wire.

Then I'll add the widths of the flags to calculate the total length along the wire of the three flags.

Communication Checklist
- ☑ Did you explain your thinking?
- ☑ Did you show how you calculated each step?
- ☑ Did you explain how you checked each answer?
- ☑ Did you show the right amount of detail?

128

A. Complete Li Ming's plan.
B. Write a solution for Carry Out the Plan. Use the Communication Checklist.

Reflecting

1. Why is it important to check each calculation in the solution of a multi-step problem?
2. Why is it important to include all steps in a written solution?

Checking

3. Akeem is painting the walls in two rooms in a Boys and Girls Club. He has two 0.90 L cans of paint and one 3.79 L can of paint. One litre of paint can cover between 23 m^2 and 37 m^2. Two walls in each room measure 6.00 m by 2.44 m. The other two walls in each room measure 4.00 m by 2.44 m. Will he have enough to paint all the walls with two coats of paint?

Akeem's Written Solution

Understand the Problem
I need to determine if I have enough paint to paint both rooms with two coats.

Make a Plan
This problem will take more than one step and more than one operation to solve.

First I need to multiply to calculate the area of the walls in each room.

Complete Akeem's written solution. Use the Communication Checklist.

Practising

4. Alexander worked with his mother at a cooperative food store. He measured the mass of two chunks of cheddar cheese for a customer. Each kilogram of cheese costs $35.00. What is the total cost of the cheese? Write a solution. Use the Communication Checklist.

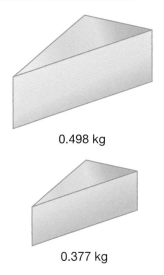

0.498 kg

0.377 kg

CHAPTER 4

Skills Bank

LESSON 1

1. Use mental math to calculate.
   a) 500 − 290
   b) 1200 − 750
   c) 600 + 1575
   d) 2499 + 501
   e) 399 + 299 + 199
   f) 5000 − 1775
   g) 5500 + 4500 + 2000
   h) 7100 − 98
   i) 2550 + 2450 + 2010

2. Explain how knowing 1000 + 2000 = 3000 can help you solve each problem.
   a) A plane flew 999 km to one city and 2001 km to another city. What is the total distance the plane flew?
   b) A container has 3000 mL of juice. 1001 mL of juice is poured out. How many millilitres are left in the container?

3. Use mental math to determine the missing number. Describe your strategy.
   a) 250 + 750 + ■ = 2000
   b) ■ + 198 = 500
   c) 1000 + ■ + 1500 = 6000
   d) 275 + 350 + ■ = 1000

LESSON 2

4. Use estimation to decide which answers are reasonable. Correct the unreasonable answers.
   a) 5679 + 2457 = 8136
   b) 987 + 304 + 723 + 815 = 1829
   c) 56 767 + 45 283 = 102 050
   d) 3045 − 283 = 2762
   e) 12 080 − 3479 = 7601
   f) 54 213 − 24 618 = 39 595

5. In 2004, as part of the Kids Can Vote project, elementary and secondary students in Canada voted online in the federal election. The chart shows the voting results for Ontario.
   a) Estimate how many students voted in the election. Show your work.
   b) Estimate the difference in votes between the Liberals and the Greens. Show your work.

Party	Votes
Conservatives	34 182
Greens	14 848
Liberals	40 735
New Democrats	31 009
Other	4658

LESSON 3

6. Estimate and calculate. Show your work.
   a) 145 + 289 + 365
   b) 763 + 201 + 437 + 675
   c) 999 + 999 + 999 + 999
   d) 876 + 459 + 888 + 555
   e) 457 + 235 + 608 + 236
   f) 910 + 192 + 367 + 843

7. Calculate the perimeter of each shape. Show your work.
   a) 256 m, 238 m
   b) 575 m, 379 m, 268 m, 854 m

8. The chart shows the number of public schools in Toronto.

Toronto Catholic		Toronto Public		
Elementary	Secondary	Elementary	Secondary	Adult
187	39	451	102	5

   a) How many more schools are in the Toronto District Public School Board than in the Toronto Catholic District School Board?
   b) Explain how you know that your answer is reasonable.

9. Calculate. Determine if your answers are reasonable. Show your work.
   a) 10 000 − 476
   b) 2345 − 906
   c) 30 000 − 5619
   d) 65 300 − 2366
   e) 35 005 − 6521
   f) 98 301 − 6725

10. An arena seats 20 000 people. The chart shows the number of people in each section.
    a) How many empty seats are in the arena?
    b) Explain how you know your answer is reasonable.

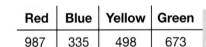

Red	Blue	Yellow	Green
987	335	498	673

11. A powwow was attended by 12 459 people. Of these people, 1086 were participants. How many people were not participants?

12. Use mental math to calculate. Describe your strategy for two of your answers.
    a) 1.79 + 2.1
    b) 3.50 + 2.25
    c) 6.79 + 2.21
    d) 6.99 + 3.25
    e) 6.00 − 2.75
    f) 10.00 − 4.99
    g) 6.25 − 1.98
    h) 4.000 − 1.999

13. Madison used a $20 bill to pay for a $1.96 drink and a $3.55 sandwich. Use mental math to calculate the change she received. Show your work.

14. Joshua's height was 1.655 m on his 12th birthday and 1.800 m on his 13th birthday. Use mental math to calculate how much he grew in that year. Show your work.

15. Calculate. Determine if your answers are reasonable. Show your work.
    a) 1.45 + 2.38
    b) 6.666 + 7.77
    c) 15.099 + 3.771
    d) 0.765 + 0.238
    e) 3.457 + 0.9
    f) 3.55 + 4.8 + 2.089

16. What is the total mass of the rice?

0.876 kg   1.248 kg

17. a) What was the total precipitation in Seymour Falls, B.C., in January and February?
    b) How much more precipitation is needed in March to equal 1 m?

Precipitation in Seymour Falls

Month	Jan	Feb
Precipitation (m)	0.516	0.447

18. Calculate. Show your work.
    a) 1.00 − 0.65
    b) 2.000 − 0.877
    c) 10.000 − 4.762
    d) 2.145 − 0.9
    e) 4.666 − 1.999
    f) 32.085 − 15.396

19. Use addition to check one answer in Question 18. Show your work.

20. The diagram shows the boundaries of a nature park. Amelia is at sign 1. She wants to see the beaver dam at sign 4.
    a) How much farther will she have to walk along the park boundaries than along the road?
    b) Are your calculations reasonable? Explain.

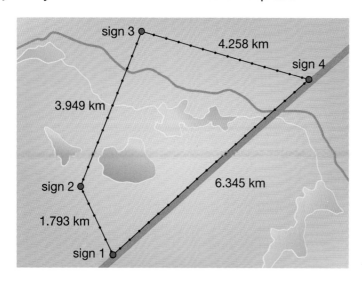

CHAPTER 4

Problem Bank

LESSON

1

1. The timeline shows some dates in Canadian history. Use the clues and mental math to identify each date. Describe your strategy.

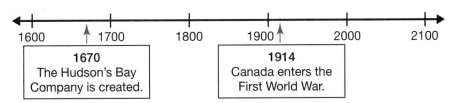

a) York becomes the capital of Upper Canada 126 years after the Hudson's Bay Company is created.
b) Canada officially becomes a country 71 years after York becomes the capital of Upper Canada.
c) The Cree, Beaver, and Chipewyan First Nations and the Government of Canada sign a treaty 15 years before Canada enters the First World War.

2

2. A recycling company is hoping to collect over 400 000 plastic milk containers by December 31 to raise money for a food bank. About how many more containers need to be collected in December to reach the goal? Explain how you estimated.

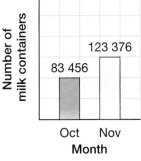

3

3. In a magic square, the numbers in the columns, rows, and diagonals have the same sum. This magic square uses the whole numbers from 351 to 366. Complete the magic square.

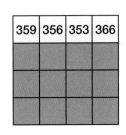

4. a) What is the greatest sum of 4 different three-digit whole numbers? Explain your reasoning.
   b) What is the least sum of 4 different three-digit whole numbers? Explain your reasoning.

5. Comet Halley is a periodic comet that is visible at regular intervals. It was visible from Earth in 1758, 1834, 1910, and again in 1986. How many more times will it be visible before the year 2500?

6. A merchant has four masses of 0.25 kg, 0.75 kg, 2.25 kg, and 6.75 kg and balance scales.
   a) What is the mass of the bag of wild rice? Show your work.
   b) Determine at least 5 other masses the merchant can measure with his four masses and balance scales. Show your work.

7. Suppose you won a $50 gift certificate for a door plaque. The chart shows the cost including taxes of each type of door plaque.

Up to 4 letters	$7.95
5 or 6 letters	$9.95
7 or 8 letters	$11.95
9 to 11 letters	$13.95
Extra letters	$0.50 each

   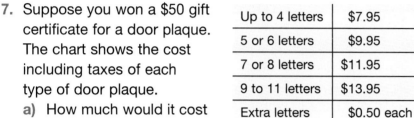

   a) How much would it cost for a door plaque for you and three of your friends?
   b) What change would you receive?

8. Logan entered a whole number into his calculator. He subtracted a decimal number and his answer ended in .475. List three numbers that he might have entered.

9. Megan measured the mass of two bags of sesame seeds. What mass of seeds should she pour from the larger bag to the smaller bag so that both bags have the same mass? Explain your reasoning.

10. Conner stopped three times to take a drink before he completed a 10.000 km cross-country ski trail. His first stop was 3.998 km from the start. His third stop was 1.200 km from the finish. His second stop was halfway between his first and third stop. What is the distance from the start to his second stop? Show your work.

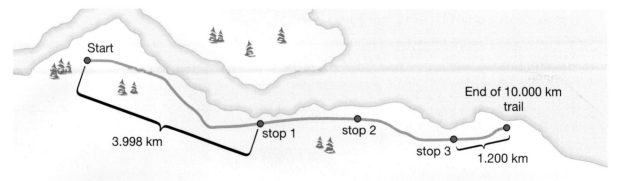

CHAPTER 4

Frequently Asked Questions

Q: How can you add two decimal numbers?

A: Depending on the numbers, you can use mental math or pencil and paper to add two decimal numbers.

Mental Math	Pencil and Paper
Change to numbers that are easier to add. $+0.002 \quad -0.002$ $1.998 + 2.856$ Add 0.002 to 1.998 and subtract 0.002 from 2.856. $1.998 + 0.002 = 2.000$ $2.856 - 0.002 = 2.854$ Now it's easy to add 2.000 to 2.854. $2.000 + 2.854 = 4.854$ The sum is 4.854.	These numbers are too hard to add using mental math. $\quad\ 3.486$ $\quad\ 2.879$ $\quad\ 5$ $\quad\ 1.2$ $\quad\ 0.15$ $\quad\ 0.015$ $\quad\ 6.365$

Q: How can you subtract two decimal numbers?

A: You can use mental math or pencil and paper to subtract two decimal numbers.

Mental Math	Pencil and Paper
Change to numbers that are easier to subtract. $+0.002 \quad +0.002$ $10.000 - 2.998$ Add 0.002 to each number. $10.000 + 0.002 = 10.002$ $2.998 + 0.002 = 3.000$ Now it's easy to subtract 3.000 from 10.002. $10.002 - 3.000 = 7.002$ The difference is 7.002.	These numbers are too hard to use mental math to subtract. Use regrouping to subtract. $\quad\ \ \ \ \ \ 9\ \ 9\ \ 16$ $\quad\ \ 12\ \ \cancel{10}\ \ \cancel{10}$ $\quad\ \cancel{1}\cancel{3}.\cancel{0}\ \cancel{0}\ \cancel{6}$ $-\ \ 6.8\ 3\ 7$ $\quad\ \ \ 6.1\ 6\ 9$

Q: When using mental math, why do you add to both numbers when subtracting but add to one number and subtract from the other when adding?

A: When you add the same number to the first and second numbers in a subtraction question, both numbers increase by the same amount. So the difference between the numbers is still the same.

When you add to a number in an addition question, the sum also increases. So you have to subtract the same amount from the other number to have the same sum.

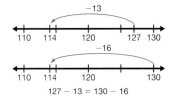

$127 - 13 = 130 - 16$

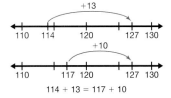

$114 + 13 = 117 + 10$

CHAPTER 4

Chapter Review

LESSON 1

1. Use mental math to calculate each answer. Describe your strategy for two of your answers.
   a) 175 + 225   b) 1999 + 1999   c) 1000 − 325   d) 2000 − 1698

2. a) A plane is flying at a height of 5000 m. It goes down 299 m. What is its height now? Use mental math. Describe your strategy.
   b) The plane then goes up 500 m. What is its height now? Use mental math. Describe your strategy.
   c) How far does it have to go down to get back to a height of 5000 m? Use mental math. Describe your strategy.

LESSON 2

3. Use estimation to decide which answers are reasonable. Show how you estimated. Correct the unreasonable answers.
   a) 66 457 + 9342 = 85 799
   b) 8765 + 5086 = 13 851
   c) 675 + 498 + 306 + 198 = 1477
   d) 10 000 − 3659 = 6341
   e) 6549 − 686 = 5863
   f) 23 000 − 8187 = 14 813

LESSON 3

4. Calculate. Determine if your answers are reasonable. Show your work.
   a) 577 + 578 + 579 + 576
   b) 105 + 295 + 457 + 961
   c) 111 + 222 + 333 + 444
   d) 499 + 499 + 499 + 499

5. What is the difference between the perimeters of the shapes? Show your work.

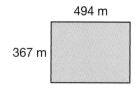
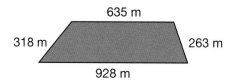

LESSON 4

6. Calculate. Show your work.
   a) 2000 − 758 = ■
   b) 6091 − ■ = 6000
   c) 18 534 − 6990 = ■
   d) 56 312 − 3626 = ■
   e) 15 500 − ■ = 2849
   f) 89 130 − 7625 = ■

7. Choose one of your answers from Question 6. Explain how you know that the answer is reasonable.

8. The population of Cobourg, Ontario, in 2001 was 17 175. The number of males was 8050. The number of girls aged 14 and under was 1510. How many females were older than 14? Show your work.

9. Use mental math to calculate each answer. Show your strategy for two of your answers.
   a) $1.75 + $10.25
   b) 1.99 + 2.28
   c) 4.998 + 1.002
   d) 20.00 − 13.97

10. Calculate. Determine if your answers are reasonable. Show your work.
    a) 4.89 + 1.50
    b) 0.7 + 0.36
    c) 3.865 + 0.95
    d) 5.005 + 1.997 + 0.998
    e) 1.589 + 2.300
    f) 10.498 + 0.532

11. The height of the cab on a truck is 2.750 m. There is a radio aerial with a height of 1.475 m on top of the cab. Will the aerial hit a highway overpass with a height of 4.000 m? Show your work.

12. Calculate. Determine if your answers are reasonable. Show your work.
    a) 5.8 − 2.5
    b) 20.06 − 18.56
    c) 0.867 − 0.099
    d) 12.058 − 3.555

13. What is the mass of the bag and its contents?

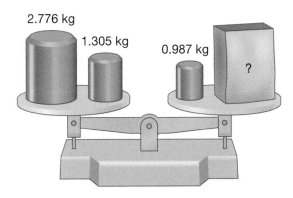

14. For a nutrition project, Craig found the mass of some fruit. Then he found the mass of their rinds or peels. What is the mass of each fruit without its rind or peel? Show your work.

Fruit	Mass of fruit (kg)	Mass of rind or peel (kg)
watermelon	1.505	0.497
banana	0.208	0.083
orange	0.195	0.064

15. Paige's school is trying to raise $2000 for UNICEF (United Nations Children's Fund). They ran a car wash over two weekends. They charged $3.00 to wash each car. In four days, they washed 87, 104, 77, and 124 cars. How much more money do they have to raise to reach their goal of $2000? Write a solution.

CHAPTER 4

Chapter Task

Gold Coins

Kurt has a collection of Canadian Maple Leaf gold coins. He mounted them on the pages of a coin collection book.

Troy ounces are often used to measure the weight of gold coins. The chart shows the troy ounces, kilograms, and value of each of the five types of Canadian Maple Leaf gold coins.

troy ounces	1.000	0.500	0.250	0.100	0.050
kilograms	0.031	0.016	0.008	0.003	0.002
value	$50	$20	$10	$5	$1

? **What is the value and mass of Kurt's gold coins?**

A. Three gold coins on one page have a total weight of 1.750 troy ounces. What is the value of the coins on that page?

B. What is the mass of the three coins in kilograms?

C. Five gold coins on another page have a total weight of 0.500 troy ounces. What is the value and mass of the five coins?

D. One page of coins has a total value of $27. How many troy ounces and kilograms of gold coins does the page have?

E. Create and solve your own problem about a gold coin collection.

Task Checklist

☑ Did you show all of your steps?

☑ Did you check the reasonableness of your calculations?

CHAPTER 5

Measuring Length

Goals

You will be able to

- choose, use, and rename metric length measurements
- measure perimeters of polygons
- solve problems using diagrams and graphs

Running in a Triathlon

CHAPTER 5

Getting Started

You will need
- a ruler or tape measure

Racing Snails

Angele timed a snail racing down the path between two rows of carrots in her garden.

? What other race tracks can you make so the snail travels the same distance?

A. How much time did it take the snail to go from one end of the row of carrots to the other?

B. The snail moved about 80 cm each minute. How many centimetres long must the row be?

C. How many metres long is the row?

START

FINISH

D. Sketch a rectangular race track that covers the same distance as your answer to Part B or C.

E. Which measurement description did you use to help you answer Part D: the one in metres or the one in centimetres? Why?

F. Sketch two other race tracks that aren't rectangles. They should cover the same distance as your answer to Part B or C. Label the side length measurements.

Do You Remember?

1. Copy and complete the relationships.
   a) 2 km = ■ m
   b) 1.4 m = ■ cm
   c) 1.4 m = ■ mm
   d) 15 cm = ■ mm

2. Calculate the perimeter of each shape.

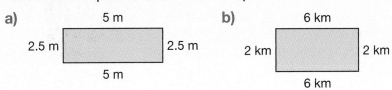

3. Sketch two different rectangles with a length of 6.7 cm and a perimeter close to, but not exactly, 18 cm. Record the width of each.

4. How much time does each event take?
   a) swim class
   b) checking your e-mail

CHAPTER 5

1 Measuring Length

Goal Select an appropriate measuring unit.

James's family went to Lighthouse Park in West Vancouver. The park has many very old and tall fir, cedar, and arbutus trees.

James tried to decide which unit he would use to measure the trip and everything he saw. He thinks he can use **decametres** to measure the length of a path around the base of this tree.

? What units of length are appropriate for describing their trip?

James's Comparison

Dad stood next to a young tree that is just about as tall as he is. Dad's height is 195 cm.

decametre (dam)
A unit of measurement for length
1 dam = 10 m

A. What unit would you use to describe the height of the older trees in the park? Explain your choice.

B. What unit would you use to describe the distance James's family travelled to get to Lighthouse Park? Why?

C. James picked up a piece of bark from a fallen arbutus tree. The bark is known for being very thin. What unit would you use to describe its thickness?

D. What unit would you use to describe the thickness of a cedar leaf?

E. What other things in a West Coast rain forest might be measured in the same units as your answer to Part D?

Reflecting

1. Why did all of the units you chose include "metre"?

2. Suppose a leaf was 7.2 cm long. Why can you say that it had been measured to the nearest millimetre?

3. Why are different units appropriate for different measurements?

Checking

4. Select an appropriate unit for each measurement. Explain each choice.
   a) the height of a room
   b) the thickness of a window
   c) the distance across your town or city

Practising

5. A nursery is selling young trees. What units are appropriate for these measurements? Explain.
   a) the heights of the trees
   b) the lengths of the leaves of the trees

6. For which measurements are centimetres appropriate? Explain.
   a) the thickness of your math book
   b) the thickness of your fingernail
   c) the thickness of 100 sheets of paper

7. Why might you measure the thickness of a wire in millimetres rather than centimetres?

8. Ryan is in a video store. What items in the store, if any, might be measured in these units?
   a) millimetres
   b) centimetres
   c) metres
   d) kilometres

9. Is it likely that a tall building would be measured in kilometres? Explain.

10. A decimetre is 0.1 metre. A decametre is 10 m. What might be the name of the unit between decametres and kilometres? Explain.

CHAPTER 5

2 Metric Relationships

Goal: Interpret and compare measurements with different units.

A group in Mexico City broke the record for the world's largest sandwich in 2004.

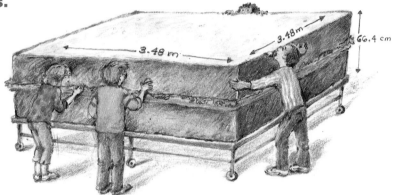

? How can you describe the measurements of the world's largest sandwich?

Denise's Description

Each piece of bread is square in shape. Each side of the square is longer than three metre sticks.

A. Why might Denise have described the bread this way?

B. Why can each side of the square be described as 348 cm long?

C. If you measured the length of the side of the square in millimetres, why might you end up with a value between 3475 mm and 3485 mm?

D. The filling of the sandwich is about 10 cm thick. Describe the thickness of each slice of bread in centimetres.

E. If you measure the thickness of the filling in millimetres, what are some of the values you might expect to measure?

F. Suppose you line up 1000 of these sandwiches and want to know the total distance. Why might you describe the sandwich as between 0.003 km and 0.004 km across?

Reflecting

1. Why might you describe 3.48 m as 348 cm, but not as 348.0 cm?

2. How do you rename each measurement?
   a) a metre description in centimetres
   b) a centimetre description in millimetres
   c) a kilometre description in metres

Checking

3. Is a slice of bread from the record sandwich taller or shorter than the height of your classroom? Explain.

4. How many millimetres tall is the sandwich?

Practising

5. Draw a picture that shows the shape of a regular slice of bread. Describe the measurements of the shape in each unit.
   a) centimetres   b) millimetres   c) metres

6. Rename each measurement using the new unit. Explain your thinking.
   a) 3.45 m to centimetres
   b) 3.4 m to centimetres
   c) 3.045 m to millimetres
   d) 2.1 km to metres

7. Liam designed this triangular flower bed.
   a) Calculate the length of the third side.
   b) If you measured the length of the 220 cm side with a millimetre ruler, could you discover that it measures 2195 mm? Explain.

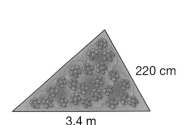
220 cm
3.4 m
perimeter is 8.5 m

8. Olivia said her pencil is 14.0 cm long. How do you know she measured to the nearest millimetre?

CHAPTER 5

3 Perimeters of Polygons

You will need
- a ruler

Goal Measure perimeters of polygons and draw polygons with given perimeters.

Qi and Raven have each designed giant floor game boards for the class fun day. They've decided to put a bright red border along the edges of the boards. Both boards require the same length of red border.

Qi's Description

My game board is a rectangle. It's 2 m wide and 3 m long.

? What could Raven's game board look like?

A. What is the perimeter of Qi's board?

B. Imagine that Raven's board is a square. Sketch this board and label the side lengths.

C. Draw the new board so that 1 cm of your picture represents 1 m of edge length on the board.

D. Suppose Raven's board is a parallelogram, but not a rectangle. Sketch a possible board and label the side lengths. Remember that the opposite sides of a parallelogram are the same length.

E. Draw the board from Part D so that 1 cm of your picture represents 1 m of board length.

F. Make a sketch to show that Raven's board could be a triangle. Label the lengths of the sides.

G. Sketch two other possible shapes for Raven's board. Label the side lengths.

Reflecting

1. Could you have drawn a different parallelogram for Part D? How do you know?

2. Why are there always many different shapes with the same perimeter?

Checking

3. a) Measure the perimeter of the triangle shown.
   b) Draw another shape with the same perimeter.

Practising

4. a) Measure the perimeter of each polygon.

    i) 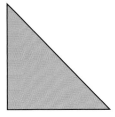 ii) 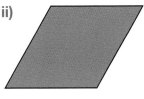 iii)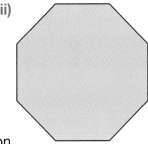

   b) Draw a polygon with the same perimeter as the octagon.

5. Draw two other shapes with the same perimeter as the parallelogram in Question 4.

6. A baker is piping icing around the edges of these cookies. Each millilitre of icing can pipe about 4 cm. How many millilitres of icing are needed for each cookie?

    a)  b) 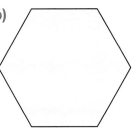 c)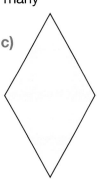

7. Why is the perimeter of the pink half of this design more than half of the perimeter of the whole design?

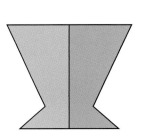

CHAPTER 5

Frequently Asked Questions

Q: How do you choose which metric unit to use to describe the length of something?

A: Use kilometres for long distances. Use metres for room-size distances. Use centimetres for relatively short distances. Use millimetres for very short lengths.

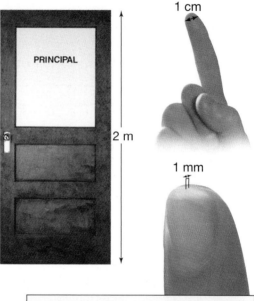

Q: How do I rename a length measurement using different metric units?

A: Use the facts in the table. For example:
3.42 m = 300 cm + 42 cm = 342 cm
       = 3000 mm + 420 mm = 3420 mm
2.1 km = 2000 m + 100 m = 2100 m
2.4 cm = 20 mm + 4 mm = 24 mm

1 km = 1000 m	1 m = 0.001 km
1 m = 100 cm	1 cm = 0.01 m
1 cm = 10 mm	1 mm = 0.1 cm

Q: How is the measurement 3.2 m different from 3.20 m?

A: A measurement of 3.2 m means that the length was measured to the nearest tenth of a metre. If the ribbon is later measured to the nearest hundredth of a metre, the length might be reported as 3.20 m, but it might also be reported as, for example, 3.18 m (since 3.18 rounds to 3.2). The length measured to the nearest hundredth of a metre might be anywhere between 3.15 m and 3.25 m.

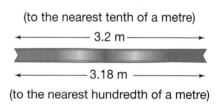

(to the nearest tenth of a metre)
3.2 m
3.18 m
(to the nearest hundredth of a metre)

Q: Can two different shapes have the same perimeter?

A: Yes. For example, all of these shapes have a perimeter of 22 cm.

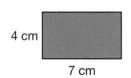
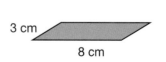
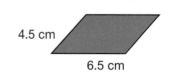
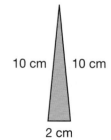

CHAPTER 5
Mid-Chapter Review

1. What unit would you use for each measurement?
   a) the height of a soup can
   b) the width of an electrical cord
   c) the distance from your home to the grocery store

2. Name a length or width you would measure in millimetres. Explain why.

3. Rename each measurement using the new unit.
   a) 2.5 cm to millimetres
   b) 6.20 m to centimetres
   c) 0.5 km to metres
   d) 4.200 m to millimetres

4. Ella reports that the height of a tree in her school yard is 3.8 m. If Ella measured the tree to the nearest hundredth of a metre, what is the least height she might report?

5. Which measurement could describe each length or distance?
   a) length of a football field
   b) height of a room
   c) width of a sidewalk

   270 cm
   1.40 m
   100 m
   50 m

6. Measure the perimeter of each shape.
   a)
   b)

7. Create a different shape with the same perimeter as the triangle in Question 6.

8. A shape has a perimeter of 3.00 m. Is it possible that one side length is only 5 cm? Explain.

CHAPTER 5

4 Solve Problems Using Logical Reasoning

You will need
- a calculator

Goal Use logical reasoning to solve a problem.

Jorge and his sister train by running on square paths around their school gyms. Jorge's path is 11.00 m longer than his sister's.

❓ How much longer are the sides of the path at Jorge's school than at his sister's school?

Marc's Chart

Understand the Problem
I know that Jorge's gym is bigger than his sister's. I also know that the running paths are the perimeters of squares and that the measurements were made to the nearest hundredth of a metre.

Make a Plan
I'll use **logical reasoning**.
I know that when the side of a square gets 1.00 m longer, then the perimeter has to get 4.00 m longer.

Carry out the Plan
I make a chart to show what happens when the side length increases.
I notice that 11.00 m is between 8.00 m and 12.00 m, so the extra side length must be between 2.00 m and 3.00 m.

I'll try an extra 2.50 m in side length.
4 × 2.50 = 10.00, so the perimeter is only 10.00 m greater.
I'll try 2.75 m longer.
4 × 2.75 = 11.00, so the sides of the path at Jorge's school must be 2.75 m longer than at his sister's school.

logical reasoning
A process for using the information you have to reach a conclusion. For example, if you know all the students in a class like ice cream and that Jane is in the class, you can logically reason that Jane likes ice cream.

Extra side length (m)	Change in perimeter (m)
1.00	4.00
2.00	8.00
3.00	12.00
???	11.00

Reflecting

1. How did Marc know the extra side length must be between 2.00 m and 3.00 m?
2. Why did Marc try a length of 2.75 m after he tried 2.50 m?
3. How did logical reasoning help Marc solve the problem?

Checking

4. When a rectangle's length is doubled and the width stays the same, its perimeter increases by 15.0 cm. How many centimetres longer is the new rectangle than the old one? Explain your reasoning.

Practising

5. Two games use cards in the shape of equilateral triangles. The card from one game has a perimeter that is 10.0 cm longer than the card from the other game. About how much shorter is the side length of the smaller card? Explain your reasoning.

6. Draw a shape with eight sides with a perimeter of 30 cm. The sides do not have to be of the same length.

7. A square porch and a porch shaped like a regular hexagon each have the same perimeter.
   a) Which shape has a longer side?
   b) Express the length of the shorter side as a fraction of the length of the longer side.

8. How many numbers between 100 and 500 have a 4 as at least one of the digits?

9. Sophia opened a book and multiplied the two page numbers she saw. The product was 5550. What was the greater page number?

Curious Math

Triangle Sides

You will need
- string
- scissors
- a ruler

Try this with a classmate.

Cut a piece of string about 35 cm long. Tie the ends together to make a loop that measures 30 cm.

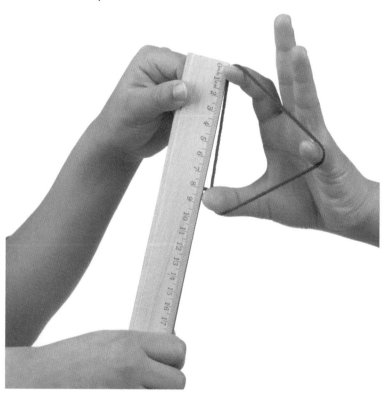

Make a triangle with the string using your fingers. Have your classmate measure and record the side lengths. Repeat to form other triangles, taking turns measuring.

1. What is the longest side that seems possible? Why might that be true?
2. How do you know that the shortest side of a triangle has to be no more than $\frac{1}{3}$ of the perimeter?
3. List all possible combinations of lengths of triangles with whole-number side lengths and a perimeter of 30 cm.

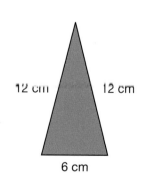

Math Game

Lines, Lines, Lines

You will need
- a die
- a ruler

Number of players: Any number

How to play: Estimate lengths in millimetres.

Step 1 Create a two-digit number by rolling a die twice. The first roll tells the tens digit. The second roll tells the ones digit.

Step 2 Without using a ruler, try to draw a line segment that is the same length as your two-digit number, measured in millimetres.

Step 3 Measure your line segment to see how close you were.
Score 2 points if you are within 5 mm.
Score 1 point if you are between 5 mm and 10 mm away.

Take turns.

Play until one player has 10 points.

Isabella's Turn

I rolled a one, then a four. I have to draw a line 14 mm long.

That's more than 1 cm but less than 2 cm.

I think it's this long.

I measure my line. It is 17 mm, so I'm 3 mm away.

I get 2 points.

CHAPTER 5

Exploring Perimeter

You will need
- a ruler
- 1 cm grid paper

Goal Explore the relationship between perimeter and area measurements.

? How can you keep increasing the perimeter of shapes that you draw inside a square?

A. Draw five squares, each with a side length of 10 cm. Calculate the perimeter and area of the squares.

B. Inside the first square, draw a polygon like the orange one in the grid.

C. Measure the perimeter and area of the new polygon. Are they greater or less than the perimeter and area of the original square?

D. In one of the other squares, make a polygon with a perimeter of 50 cm. Use the other squares to make shapes of perimeter 70 cm, 80 cm, and 100 cm. Each time, calculate the area of your shape.

E. How did the areas change?

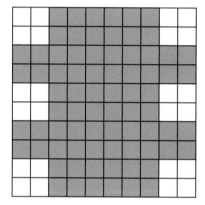

Reflecting

1. Does a polygon with more sides always have a greater perimeter? Explain using examples from class work.

2. What strategies did you use to make sure the perimeter was the required value?

3. Rebecca says she has a polygon with a perimeter of 80 cm. Can you predict the area of Rebecca's polygon? Explain using examples.

Mental Math

Calculating Lengths of Time

You can use different methods to calculate how much time something took.

Akeem's Method

I can add on to calculate the difference between the two times.

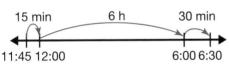

The difference is 6 h 45 min.

Li Ming's Method

I can add on too much time and then subtract to calculate the difference between the two times.

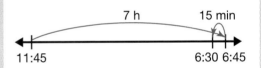

The difference is 6 h 45 min.

A. Explain Akeem's method.

B. Explain Li Ming's method.

Try These

1. Calculate the difference between the two times.

   a)

   c)

   b)

   d)

CHAPTER 5

Skills Bank

LESSON 1

1. Select the most appropriate unit to measure each length or distance.
   a) the length of a paper clip
   b) the thickness of a drum hide
   c) the height of a lamp
   d) the distance around the school playground

2. Name two distances or lengths you might measure with each of these units.
   a) millimetres
   b) centimetres
   c) metres
   d) kilometres

LESSON 2

3. Rename each measurement using the new unit.
   a) 3.42 m to centimetres
   b) 5.213 km to metres
   c) 0.25 m to millimetres
   d) 13.2 cm to millimetres

4. The first column describes a measurement that was rounded to the nearest whole centimetre. Fill in the other columns to describe what the measurement might have been if it had been rounded to the nearest tenth of a centimetre.

	Rounded measurement	Lowest possible value	Highest possible value
a)	34 cm	■ cm	■ cm
b)	89 cm	■ mm	■ mm

5. The first column describes a measurement that was rounded to the nearest tenth of a kilometre. Fill in the other columns to describe what the measurement might have been if it had been rounded to the nearest metre.

	Rounded measurement	Lowest possible value	Highest possible value
a)	2.1 km	■ m	■ m
b)	3.4 km	■ m	■ m

6. Draw a four-sided polygon with each perimeter.
   a) 26 cm
   b) 142 mm
   c) 8.7 cm

7. Measure the perimeter of each polygon.
   a)

   c)

   b)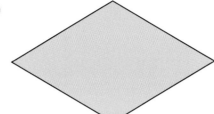

   d)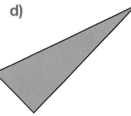

8. What is the perimeter of each polygon inside the square?
   a)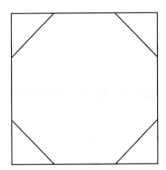

   b)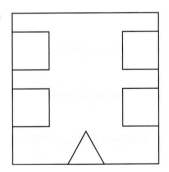

   c)

9. a) Draw two rectangles with the same perimeter but different areas.
   b) Draw two polygons with the same area but different perimeters.

CHAPTER 5
Problem Bank

LESSON

2
1. A 50 g ball of yarn is about 125 m long. What is the mass of a ball of yarn 2 km long?

3
2. The perimeter of a square is 396 m. What is the length of the sides?

3. Draw two non-congruent 5-sided polygons. Use these clues:
   - The perimeter is 40 cm.
   - There are at least two equal sides.
   - At least one side has a length of 10 cm.

4
4. Draw a 10 cm by 12 cm rectangle. Use these clues to draw 5 different shapes inside it:
   - The shapes do not overlap.
   - The total perimeter for all 5 shapes is greater than 60 cm.

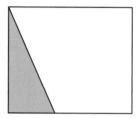

5. The perimeters of two squares differ by 8 cm. The total perimeter for the two squares is 96 cm. What is the side length of the larger square?

5
6. A rectangle has whole number side lengths. The length of the rectangle is tripled and the width is doubled. The new area is 204 cm^2. The new perimeter is 110 cm. What was the old perimeter?

7. You can make a "snowflake" polygon from an equilateral triangle. At every step, the middle third of each side is bumped out in the shape of a smaller equilateral triangle. Calculate the perimeter of the polygons shown for steps 1 to 3.

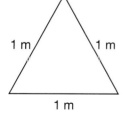
step 1

step 2

step 3

CHAPTER 5

Frequently Asked Questions

Q: When might you solve a problem by using logical reasoning?

A: When you have some information that you think you could use to figure out something else, you might use logical reasoning.

For example, suppose that you know that the length of a rectangular room is 3 times the width and that the perimeter is 48 m. You want to determine the length of the room.

You can use logical reasoning:
- Since opposite sides of a rectangle are equal, the length plus the width is the same as half the perimeter.
- Half the perimeter is 24 m.
- Since the length is three times the width, adding the length and width is like adding four times the width.
- That means four times the width is 24 m.
- The width must be 6 m, so the length must be 18 m.

Q: Why is it possible to create polygons with a greater perimeter, but a similar area?

A: Area and perimeter are different measurements. For example, it you took the green rectangle below and cut it in half, you could make the yellow rectangle. You'd have the same area, but a much greater perimeter.

CHAPTER 5

Chapter Review

1. Why would you measure the total length of a bus in metres rather than centimetres?

2. What unit would you use to measure the thickness of a cup? Why?

3. Rename the measurements Owen made of his keyboard so that they are all in centimetres.

4. In what ways is the measurement 3.2 m like 3.20 m? How is it different?

5. Arden spilled maple syrup on her notes. What measurement units make the sentence true?

3.64 ● is the same as 364 ●. If I measure in ● it might be any length between 3635 ● and 3645 ●.

6. Draw a square and a triangle that each have the same perimeter as this rectangle.

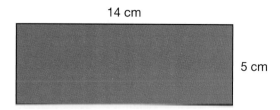

7. Draw a rectangle with a perimeter that is 6 cm more than the perimeter of the rectangle in Question 6.

8. A square and an equilateral triangle each have a perimeter of 72 cm. How much longer is one side of the triangle than one side of the square?

9. A large rectangular room is 12.34 m long. What increase in width will make the perimeter 5 m longer if the length stays the same?

10. The short side of an isosceles triangle is 16 cm long. Increasing each side length by 24 cm doubles the perimeter. How long are the other sides of the original triangle? Explain your reasoning.

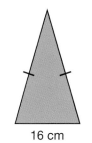
16 cm

11. One side of an isosceles triangle is 12 m longer than the other two sides. The perimeter is 84 m. How long are the sides of the triangle?

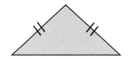

12.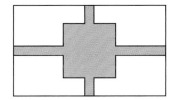

   a) Which of the two green polygons has a greater perimeter? How do you know?
   b) Which of the two has a greater area? How do you know?
   c) Does a polygon with a greater area always have a greater perimeter? Explain.

13. Draw a polygon with a perimeter of 50 cm inside a rectangle with a perimeter of 30 cm.

14. Start with a square. Suppose you cut it up and rearrange the pieces into a thin rectangle. Is the perimeter of the rectangle greater or less than the perimeter of the square? Why?

CHAPTER 5

Chapter Task

Mapping out a Biathlon Course

A triathlon involves swimming, biking, and running.

In youth triathlons, the 11- and 12-year-olds might swim 150 m to 300 m, bike 8 km to 12 km, and run 1.5 km to 3 km.

You are planning a cycling and running youth biathlon for Grade 5 and 6 students.

? How might you design courses for a youth biathlon?

A. Choose a length for each event of your biathlon.

B. Would most students be able to complete your biathlon in 40 min? Use the winning times for each event to help decide.

Some winning times:
 Cycling 18 km in 1 h
 Running 4 km in 21 min

C. Redesign your biathlon so that most students can complete it in 40 min.

D. Sketch a polygon course for each event. Each course should be on or near your school grounds.

E. Make a scale drawing that shows your biathlon courses. Label the side lengths and perimeter of each polygon.

Task Checklist
☑ Did you explain all of your calculations?
☑ Were you reasonably accurate in your measurements?
☑ Did you show all your steps?
☑ Is your scale drawing clear?

CHAPTER 6
Multiplication and Division

Goals

You will be able to

- identify prime numbers, composite numbers, factors, and multiples
- calculate products and quotients using mental math
- multiply and divide by two-digit numbers
- pose and solve problems using multiplication and division
- investigate the effects of the order of operations

Planting trees

CHAPTER 6

Getting Started

You will need
- base ten blocks
- a place value chart

Recycling Milk Cartons

A dairy company offers a coupon for every six milk cartons that are recycled. Each coupon is worth 50¢.

Each of the 72 Grade 6 students in a school each recycled 12 milk cartons. All the coupons received by the students were donated to the Food Bank.

? **What is the total value of the coupons received by the students?**

A. How many cartons did the students recycle? Show your work.

B. How many coupons did they donate to the Food Bank? Show your work.

C. Explain how you know that your answers to Part A and Part B are reasonable.

D. What is the total value of the coupons donated to the Food Bank?

E. Explain how you might use mental math to answer Part D.

F. Create and solve your own problem about receiving coupons for recycling containers.

Do You Remember?

1. Estimate. Show your work.
   a) 37 × 49
   b) 82 × 91
   c) 8)$\overline{1687}$
   d) 3)$\overline{1497}$

2. Calculate. Use mental math.
   a) 60 × 40 = ■
   b) 12 × 1000 = ■
   c) 1200 ÷ 4 = ■
   d) 4800 ÷ 6 = ■

3. Calculate.
   a) 45 × 53
   b) 96 × 28
   c) 8)$\overline{5318}$
   d) 7)$\overline{6539}$

4. In the Big Bike for Stroke campaign, each rider on a team must raise donations to ride on a big bike with 30 seats. All funds go to the Heart and Stroke Foundation of Canada. One half of the riders on a team raised $50 each. The other half raised $75 each. How much money did the team raise?

CHAPTER 6

1 Identifying Factors, Primes, and Composites

You will need
- a calculator
- a spinner
- a paper clip

Goal Identify the factors of prime and composite numbers

factor
A whole number that divides another whole number without a remainder
$8 \div 2 = 4$
2 is a factor of 8 because 2 divides 8 without a remainder.
The other factors of 8 are 1, 4, and 8.

Denise and Rodrigo are playing the game Spin and Factor. They take turns spinning two-digit numbers. The digit from the first spin represents the number of tens. The digit from the second spin represents the number of ones.

The player who spins a number expresses the number as a product of **factors**. The player's score is the total number of factors in the expression. The factors 1 and the number itself can't be used in the product.

The winner is the player with the most points after 10 spins each.

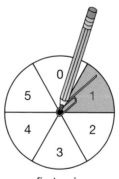
first spin

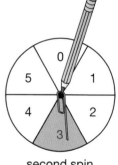
second spin

? Which number scores the greatest number of points in Spin and Factor?

Rodrigo's Solution

If I spin 13 I can't score any points. The only factors of 13 are 1 and 13, so the only product I could write is 1×13. The game rules say I don't score any points for 1×13.

I won't get points for spinning any **prime number**.

I need to spin **composite numbers** like 12 to score points.

I can express 12 as the products 1×12, 2×6, or 3×4.

I would score 2 points for 2×6 or 3×4.

prime number
A number that has only two different factors: 1 and itself

2 is a prime number because it has only two factors: 1 and 2.

composite number
A number that has more than two different factors

4 is a composite number because it has more than 2 factors: 1, 2, and 4.

Denise's Solution

I can score more points if I express a composite number using only prime numbers. I can express 12 as 4 × 3 and score 2 points. But I can also write 4 as 2 × 2. I will express 12 as 2 × 2 × 3. This scores 3 points.

16 is a better number to spin than 12. I can score 4 points for 16.

Communication Tip
The numbers 0 and 1 are neither composite nor prime.

A. How can you use Rodrigo's expressions 1 × 12, 2 × 6, and 3 × 4 to determine all the factors of 12?

B. Show how you can score 4 points after spinning 16.

C. Play Spin and Factor. Identify the numbers you spin as prime or composite. Record the factors of each composite number and how many points you scored.

D. Is it possible to score 6 points? How do you know?

E. Which number scores the greatest number of points?

Reflecting

1. How can you tell without any calculations whether a number ending in 5 is a prime or composite number?

2. How many even prime numbers are there? Explain.

3. Suppose you could spin the number 37. How can you identify 37 as a prime number or a composite number?

4. Explain your method of identifying all of the factors of a composite number. Use an example.

Checking

4. Suppose you play the game Spin and Factor using this spinner.
   a) Identify one prime number you can spin. Explain why it is a prime number.
   b) Which number scores the greatest number of points? Show your work.

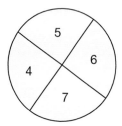

Practising

5. Identify each as prime or composite. List the factors of each number.
   a) 5
   b) 7
   c) 9
   d) 14
   e) 20
   f) 100

6. Express one composite number from Question 5 as a product of only prime numbers.

7. Jennifer expressed 81 as 3 × 27. Show how she can identify factors of 27 so that the numbers in her expression are only prime numbers.

8. Can prime numbers other than 2 and 5 ever be 3 apart? Explain.

9. Show that a number that can be written as the product of four factors can also be written as the product of three factors or two factors.

10. A two-digit number is prime. When you reverse the digits, that number is also prime. What could the number be? List three possibilities.

Curious Math

Separating Primes from Composites

You will need
- a 100 chart

You can use a 100 chart to separate prime numbers from composite numbers. This method was invented by the Greek mathematician Eratosthenes.

1. Circle 2, the first prime number, on a 100 chart. Explain how you know that 2 is a prime number.

2. Cross off every multiple of 2 after 2. How do you know that each number crossed off is not a prime number?

1	②	3	̶4̶	5	̶6̶	7	̶8̶	9	̶1̶0̶
11	̶1̶2̶	13	̶1̶4̶	15	̶1̶6̶	17	̶1̶8̶	19	̶2̶0̶

3. Circle 3, the next prime number. Explain how you know that 3 is a prime number.

4. Cross off every multiple of 3 after 3. How do you know that each number crossed off is not a prime number?

5. Repeat circling prime numbers and crossing off their multiples. What do you notice about the numbers that are not crossed off?

CHAPTER 6

2 Identifying Multiples

You will need
- centimetre grid paper
- a calculator

Goal Solve problems by identifying multiples of whole numbers.

Akeem and Isabella are studying comets. The comets Kojima and Kowal-Mrkos were both visible in 2000. Kojima appears about every 7 years. Kowal-Mrkos appears about every 9 years.

? When is the next year that both comets will be visible?

Isabella's Method

multiple
A number that is the product of two factors
8 is a multiple of 2 because $2 \times 4 = 8$

I can solve the problem by identifying the **multiples** of 7 and 9.
I'll keep multiplying 7 and 9 by counting numbers to create lists of multiples.
I'll look for a multiple that is in both lists and I'll add that number to 2000.

Multiples of 7	1 x 7 7	2 x 7 14	3 x 7 21
Multiples of 9	1 x 9 9	2 x 9 18	3 x 9 27

Akeem's Method

If I keep adding 7, I'm adding multiples of 7 to 2000.
The first multiples are 7, 14, and 21.
I'll keep adding 7s or 9s on number lines.
I'll write the years on the number lines.
I'll look for a year that is on both number lines.

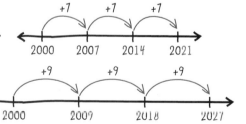

A. What will Isabella do to determine the next multiples of 7 and 9?

B. What will Akeem do to determine the next year that each comet will appear?

170

C. Identify the next 10 multiples of 7 and the next 10 multiples of 9.

D. Determine a year when both comets will be visible.

> ### Reflecting
>
> 1. How do you know that 7 must be a factor of each multiple of 7?
> 2. Is it possible to list all of the multiples of 9? Explain.

Checking

3. In 2001, the periodic comets Helin and Mueller 4 were both visible. Helin appears about every 14 years. Mueller 4 appears about every 9 years.
   a) List the first 10 multiples of 14.
   b) List the first 15 multiples of 9.
   c) What is the the next year the two comets will be visible?

Practising

4. List five multiples of each number.
   a) 2   b) 3   c) 12   d) 15   e) 25   f) 100

5. In Abigail's school, a room has 30 lockers numbered from 1 to 30. There are running shoes in each locker with a number that is a multiple of 5. There are rubber boots in each locker with a number that is a multiple of 3.
   a) Which lockers have running shoes?
   b) Which lockers have rubber boots?
   c) Which lockers have both running shoes and rubber boots?

6. Show why the second number is a multiple of the first number.
   a) 2   240   b) 3   39   c) 4   100   d) 7   714

7. The Winter Olympics are usually held every 4 years. In 2002, they were held in Salt Lake City, Utah, U.S.A.
   a) When will the 10th Winter Olympics after 2002 be held?
   b) Will there be Winter Olympics in the year 2062? Explain.

8. During the month of April, Greg dried dishes every third day and cleaned up his room every fifth day. On what days did he do both chores?

CHAPTER 6

3 Calculating Coin Values

You will need
- play money

Goal Use the relationships between coin values to simplify calculations.

Qi has a Canadian coin collection.

? What is the value in cents of his coins?

A. Qi has all 12 quarters made to celebrate Canada's 125th birthday in 1992. Why can you express the value of the quarters as 12 × 25?

B. What are some ways to calculate the total value in cents of the quarters?

C. Qi also has 16 millennium quarters. Show how to arrange his quarters to help calculate 16 × 25.

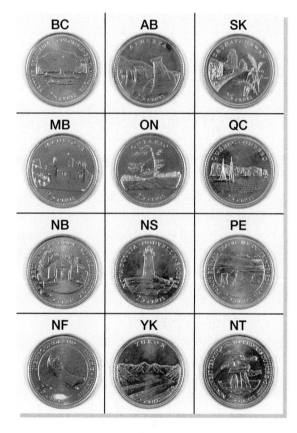

D. The last part of Qi's collection is 24 half dollars. Show how to arrange his half dollars to calculate 24 × 50.

E. Create a problem about calculating the value of a coin collection made up of nickels, dimes, quarters, or half dollars. The number of coins should be greater than 30.

Reflecting

1. How could you use money to calculate 34 × 5?
2. How can the answer in Part C help you calculate 17 × 25?
3. Choose a number greater than 20. What are some ways to use mental math to multiply that number by 5, 25, and 50?

Mental Math

Halving and Doubling to Multiply

You can multiply some numbers by halving one number while doubling the other.

To mentally calculate 8 × 15, use one half of 8 and double 15

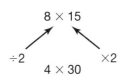

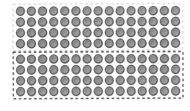

8 × 15 = 120   4 × 30 = 120

A. Why does halving one number and doubling the other number not change the product?

Try These

1. Calculate. Use the halving and doubling method.
   a) 12 × 15
   b) 12 × 150
   c) 25 × 24
   d) 22 × 5
   e) 36 × 50
   f) 50 × 22

173

CHAPTER 6

4 Multiplying by Hundreds

You will need
- a place value chart
- counters

Goal Use multiplication facts and regrouping to multiply by hundreds.

Each of Tara's digital photos contains a different number of pixels arranged in an array. Each pixel has one colour. In the 4 × 6 array, or 4-by-6 image, you can actually see the 24 pixels. In the 10-by-15 image, it is more difficult to see and count the 150 pixels. In the other photos, you cannot see or count the pixels.

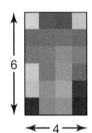 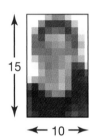 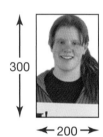  

? How many pixels are in each image?

Tara's Calculation

There are 200 rows of 300 pixels in a 200-by-300 image. To calculate the number of pixels, I need to calculate 200 × 300. I first represent 300 on a place value chart.

To multiply any number by 200, I can multiply by 100 first and then multiply the product by 2. To multiply 300 by 100 on a place value chart I move the 3 counters two columns to the left.

To multiply that product by 2, I double the number of counters.

Thousands			Ones		
Hundreds	Tens	Ones	Hundreds	Tens	Ones
			●●●		

Thousands			Ones		
Hundreds	Tens	Ones	Hundreds	Tens	Ones
	●●●		●●●		

Thousands			Ones		
Hundreds	Tens	Ones	Hundreds	Tens	Ones
	●●●●●●				

A. Why does moving the 3 counters two columns to the left represent multiplying 300 by 100?

B. What is the product of 300 and 100? Explain.

C. Why did Tara double the number of counters?

D. How many pixels are in a 200-by-300 image?

E. Show how to use a place value chart to calculate the number of pixels in the last two photos.

Reflecting

1. Why is multiplying a number by 200 the same as multiplying it by 100 and then by 2?

2. What happens to the place value of each digit when you multiply by 100? Use 100 × 400 to explain your answer.

3. How can you use mental math to multiply numbers in the hundreds? Use 400 × 300 to explain your answer.

Checking

4. Use mental multiplication to identify the image on the right with the greater number of pixels. Explain your reasoning.

700 x 800 pixels

Practising

5. Calculate.
   a) 40 × 50
   b) 40 × 100
   c) 100 × 300
   d) 100 × 500
   e) 600 × 900
   f) 5000 × 60

6. Which amount is more money? Explain.

900 x 600 pixels

7. A ream of paper contains 500 sheets. How many sheets of paper are there in 60 reams?

8. A 600-by-■ image contains 120 000 pixels. What is the value of ■? Show your work.

CHAPTER 6

5 Estimating Products

You will need
- a calculator

Goal Estimate to check the reasonableness of a calculation.

Kurt learned these facts when writing a report on energy resources in Canada.

- Giant trucks are used to carry the tar sands from the Athabasca Tar Sands region. The tar in the sand is used to make oil and other petroleum products.
- Each truck can carry 363 **tonnes (t)** of tar sand.
- It takes 2 tonnes of tar sand to make one barrel or 159 L of oil.

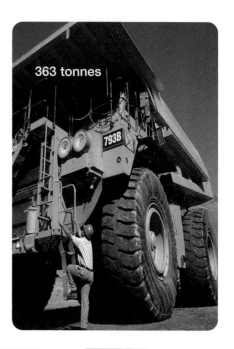

363 tonnes

? About how many litres of oil can be made from each truckload of tar sand?

Kurt's Estimation

I used mental math to estimate that about 180 barrels of oil can be made from each truckload of tar sand.

I will use my calculator to multiply 180 by 159 to estimate the number of litres of oil.

I estimate that the product of 180 and 159 will be about 18 000 + 9000 = 27 000.

tonne (t)
A unit of measurement for mass
1 t = 1000 kg

A. How can you use mental math to estimate that a truckload of tar sand makes about 180 barrels of oil?

B. Why would you multiply 180 by 159 to estimate the number of litres of oil that can be made from each truckload of tar sand?

C. How might Kurt have estimated the product of 180 and 159?

D. Use your calculator to multiply 180 by 159.

E. Was Kurt's estimate reasonable? Explain.

Reflecting

1. Explain why the answer in Part D is an estimate rather than an exact answer.

2. When you are solving a problem that has many steps, why is it a good idea to check the reasonableness of the answer for each step?

Checking

3. A mine located north of Fort McMurray, Alberta, uses 23 giant trucks to carry tar sand.
   a) Calculate the number of barrels of oil that can be made from 23 truckloads. Show your work.
   b) Show how to use estimation to check your answer.

Practising

4. Check if each answer is reasonable. Use estimation.
   a) 49 × 48 = 2352
   b) 59 × 415 = 24 485
   c) 805 × 75 = 70 375
   d) 109 × 98 = 10 682
   e) 55 × 1235 = 67 925
   f) 78 × 2485 = 293 830

5. a) Calculate the number of pixels in each image.
   b) Show that each answer is reasonable. Use estimation.

250 x 400 pixels

480 x 640 pixels

6. a) Calculate the number of hours in 1 year.
   b) Calculate the number of seconds in 1 day.
   c) Show that your answers are reasonable. Use estimation.

7. a) Describe a method to use multiplication to estimate the number of names in the white pages of your local phone book.
   b) Use your method to estimate the number of names.

CHAPTER 6

6 Multiplying by Two-Digit Numbers

You will need
- a calculator

Goal Use pencil and paper to multiply a whole number by a two-digit number.

On September 26, 2002, Jeff Adams of Toronto climbed 1776 steps of the CN Tower in a wheelchair.

? How many centimetres high did Jeff Adams climb up in his wheelchair?

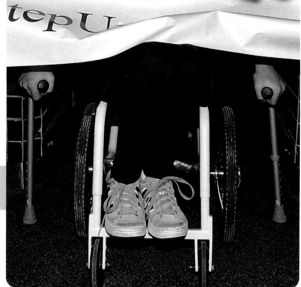

Chandra's Multiplication

The height of a step in my school is 19 cm.
The steps in the CN Tower are probably the same height.

There are 1776 steps in the tower, so I need to multiply 1776 by 19.

I estimate that the product will be around 40 000.

I first multiply 1776 by 9.

Now I need to multiply 1776 by 10 and add both products.

```
  6 6 5
  1776
×   19
 15 984
```

Maggie's Multiplication

I estimate that 1776 × 19 is about 34 000. Calculating the product of two numbers is like calculating the area of a rectangle.

I can calculate 1776 × 19 by placing the numbers in a 4-by-2 array, so I can calculate the area of the rectangle in parts. I renamed each number. Then I multiplied to complete the rows and columns in the array.

1776 × 19 = ■

	1000 +	700 +	70 +	6
10 +	10 000	7000	700	60
9	9000	6300	630	54

To calculate the product, I added the numbers in the rows and columns.

1776 × 19 = ■

	1000 +	700 +	70 +	6	
10 +	10 000	7000	700	60	17 760
9	9000	6300	630	54	15 984
	19 000	13 300	1330	114	

A. Complete Chandra's multiplication.

B. Complete Maggie's multiplication.

C. How high did Jeff Adams climb in his wheel chair?

Reflecting

1. Explain how Chandra and Maggie each might have estimated the product of 1776 and 19.

2. Is your answer to Part C more than 300 metres? Explain how you know.

3. How is multiplying a two-digit number by a four-digit number similar to multiplying a two-digit number by a two-digit number? How is it different?

Checking

4. On July 23, 1999, Ashrita Furman of New York State, jumped up 1899 steps of the CN Tower on a pogo stick.
   a) If the height of each step is 19 cm, how many centimetres did he climb on his pogo stick? Show your work.
   b) Explain how you know that your answer is reasonable.

Practising

5. Calculate. Estimate to show that each answer is reasonable.
   a) 23 × 145 = ■
   b) 75 × 246 = ■
   c) 56 × 3821 = ■
   d) 67 × 9325 = ■

6. To improve his fitness, Charles climbed 16 flights of steps in a local building. Each flight has 11 steps. The height of each step is 18 cm.
   a) Calculate the number of centimetres he climbed to reach the top of the stairs.
   b) Did he climb more than 30 m? Explain how you know.
   c) Explain how you know that each calculation is reasonable.

7. Mary's school collected $2456 in $2-dollar coins to donate to a charity. Mary decided to calculate the length in millimetres if the coins were placed end to end.
   a) How many $2 coins did her school collect?
   b) What is the length in millimetres of all the two-dollar coins placed end to end? Show your work.
   c) Is the length greater than 30 m? Explain how you know.
   d) Explain how you know that each calculation is reasonable.

8. a) Create a problem that can be solved by multiplying a two-digit number by a three-digit or four-digit number.
   b) Solve your problem.
   c) Explain how you know that your answer is reasonable.

Curious Math

Lattice Multiplication

You can often see a lattice or a frame made of crossed wood strips in gardens.

You can use a lattice to multiply two numbers such as 1638 and 24.

First, draw a 2-by-4 lattice and write 1638 on top and 24 on the right side.

Then write the product of each pair of digits inside the lattice. The diagonal line in each cell is used to separate tens and ones. For example, 6 × 4 = 24, and 24 is 2 tens and 4 ones.

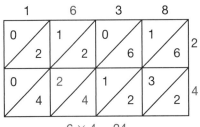

6 × 4 = 24
24 is 2 tens and 4 ones

Finally add the numbers along the diagonals, starting at the bottom right. Write any digit representing the number of tens in a sum in the next diagonal.

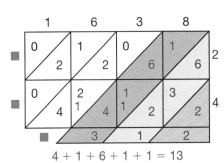

4 + 1 + 6 + 1 + 1 = 13

1. Complete the calculation to show that the product is 39 312.

2. Calculate each product. Use lattice multiplication.
   a) 41 × 65
   b) 269 × 76
   c) 2062 × 21

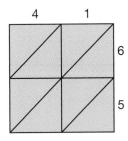
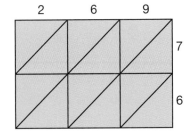
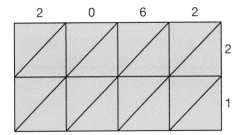

3. Calculate any product using lattice multiplication.

CHAPTER 6

Frequently Asked Questions

Q: What are factors and multiples?

A: A factor is a whole number that divides another number without leaving a remainder

For example, 5 is a factor of 60 because $60 \div 5 = 12$.
From this you can tell that 12 is also a factor of 60, because $60 \div 12 = 5$
The factors of 60 are
1, 2, 3, 4, 5, 6, 10, 12, 15, 20, 30, and 60.

A multiple is a number that is the product of two factors. For example, 60 is a multiple of 5 because $5 \times 12 = 60$.
5, 10, 15, 20, 25, ... are multiples of 5.

Q: How do I multiply by tens and hundreds?

A: You can think of place value.

For example, 200×500 can be calculated by multiplying 500 by 100 first.

$100 \times 500 = 50\,000$ because the digit 5 increases from hundreds to ten thousands.

To multiply 500 by 100 on a place value chart, move the five counters two columns to the left.

So $200 \times 500 = 2 \times 50\,000$
$= 2 \times 50$ thousand
$= 100$ thousand
$= 100\,000$.

To multiply 50 000 by 2 on a place value chart, multiply the number of counters by 2 and regroup if necessary.

182

CHAPTER 6
Mid-Chapter Review

LESSON 1

1. Identify the factors of each number.
   a) 18   b) 22   c) 36   d) 250

2. Identify each number as prime or composite. Show your work for one number.
   a) 14   b) 17   c) 91   d) 200

3. Cassandra expressed 84 as a product of four prime numbers.
   84 = ■ × ■ × ■ × ■
   Identify the prime numbers.

LESSON 2

4. a) Explain why 168 is a multiple of 7.
   b) Use your answer to Part a) to show that 7 is a factor of 168.

5. Identify three multiples of each number.
   a) 4   b) 22   c) 30   d) 150

LESSON 3

6. Calculate the value in cents or dollars of 16 quarters.

LESSON 4

7. Calculate each product. Use mental math.
   a) 40 × 60   c) 25 × 200
   b) 100 × 100   d) 500 × 600

8. A map has a scale where 1 cm represents 300 km. Calculate the number of kilometres represented by 30 cm. Use mental math.

LESSON 5

9. Which answers are reasonable? Use estimation.
   a) 45 × 798 = 35 910   c) 61 × 6105 = 272 405
   b) 98 × 135 = 23 230   d) 2077 × 33 = 68 541

10. A building contains 24 windows. Each window measures 55 cm by 95 cm.
    a) Calculate the total area of all the windows in the building.
    b) Show that each calculation is reasonable. Use estimation.

LESSON 6

11. Calculate. Estimate to show that each answer is reasonable.
    a) 25 × 48 = ■   c) 88 × 2999 = ■
    b) 44 × 385 = ■   d) 4786 × 76 = ■

12. A typical pencil could be used to draw a line about 55 km long. Each of the 279 students in George's school has five pencils.
    a) How long a line could all these pencils draw?
    b) Estimate to show that each calculation is reasonable.

CHAPTER 6

7 Dividing by 1000 and 10 000

You will need
- a place value chart
- a calculator

Goal Use mental math to divide whole numbers by 1000 and 10 000.

Every 1000th electronic appliance made by a company contains a gift coupon. Every 10 000th appliance also contains free headphones. The table shows the approximate number of each appliance made.

Appliances Made

Appliance	Number
hand-held game systems	9 500 000
MP3 players	4 500 000
DVD players	500 000
plasma TVs	60 000

? How many coupons and headphones were given away with each kind of appliance?

Rebecca's Solution

I will divide 9 500 000 by 1000 to determine the number of gift coupons for the hand-held game system.
I write 9 500 000 in a place value chart.

Millions			Thousands			Ones		
Hundreds	Tens	Ones	Hundreds	Tens	Ones	Hundreds	Tens	Ones
		9	5	0	0	0	0	0

I first divide by 10 by moving each digit one place value to the right. Each time I move the digits to the right, each value becomes 10 times smaller.
9 500 000 ÷ 10 = 950 000

Thousands			Ones		
Hundreds	Tens	Ones	Hundreds	Tens	Ones
9	5	0	0	0	0

I divide 950 000 by 10 and 9 500 000 by 100 by moving the digits another place value to the right.
950 000 ÷ 10 = 95 000; 9 500 000 ÷ 100 = 95 000

Thousands			Ones		
Hundreds	Tens	Ones	Hundreds	Tens	Ones
	9	5	0	0	0

A. Complete Rebecca's method to divide 9 500 000 by 1000.

B. How can you use your answer from part A to divide 9 500 000 by 10 000?

C. Calculate the number of gift coupons and the number of headphones for each appliance. Show your work.

Reflecting

1. Explain why you can divide a number by 100 by dividing the number by 10, two times.

2. a) How many times should you divide by 10 to divide a number by 1000? Explain.
   b) How many times should you divide by 10 to divide a number by 10 000? Explain.

Checking

3. Every 1000th music CD contains a gift coupon. Every 10 000th CD also contains a ticket to a Digerate 2 concert. The chart shows the approximate number of each CD made. Calculate the number of coupons given away with each CD. Use mental math.

CDs Made

Name of CD	Number made
The Digerate 2: Flee Australia	1 800 000
Metal Spoke	430 000
Three-Shop Pam	2 450 000
Shout Till Your Knees Hurt	60 000

Practising

4. Calculate. Use mental math.
   a) 202 000 ÷ 1000 = ■
   b) 99 000 ÷ 1000 = ■
   c) 10 000)60 000
   d) 10 000)150 000
   e) 6 000 000 ÷ 1000
   f) 6 500 000 ÷ 10 000

5. Calculate the first quotient. Show how to use the first quotient to calculate the second quotient.
   a) 1500 ÷ 10 = ■   1500 ÷ 100 = ■
   b) 15 000 ÷ 100 = ■   15 000 ÷ 1000 = ■
   c) 1 500 000 ÷ 1000 = ■   1 500 000 ÷ 10 000 = ■

6. Explain how you can divide by 1000 to estimate each quotient.
   a) 3943 ÷ 998   b) 56 012 ÷ 990   c) 988)296 482

7. A company will donate $200 for every $1000 raised by a charity. How much money would the company donate if the charity raised $48 000?

8. A globe of the Earth has a scale of 1 cm represents 1000 km. The distance around the equator is about 40 000 km. What is the distance in centimetres around the equator of the globe?

CHAPTER 6

8 Dividing by Tens and Hundreds

You will need
- a place value chart
- a calculator

Goal Use renaming and a division fact to divide by tens and hundreds.

A recent study found that about 1 message in every 200 e-mail messages is infected with a certain virus.

Each of the 33 students in Khaled's class receives about 5 e-mail messages daily.

? **How many of the class's e-mail messages are likely to be infected in one year?**

Khaled's Solution

I used a calculator to calculate that the class receives a total of 60 225 or about 60 000 messages each year.

If 1 message in 200 messages is infected, I will divide by 200 to estimate the number of infected messages.

Step 1 I write 6 in the ten thousands column to represent 60 000 on a place value chart.

Thousands			Ones		
Hundreds	Tens	Ones	Hundreds	Tens	Ones
	6				

Step 2 I can use the place value chart to rename 60 000.
60 000 = 6 ten thousands
60 000 = 60 thousands
60 000 = 600 hundreds

Thousands			Ones		
Hundreds	Tens	Ones	Hundreds	Tens	Ones
	6	60	600		

Step 3 I also rename 200 as 2 hundreds so I can divide 600 hundreds.
Dividing 600 hundreds by 2 hundreds is the same as dividing 600 by 2.
About 300 e-mails will be infected each year.

60 000 ÷ 200 = ■
600 hundreds ÷ 2 hundreds = ■
600 ÷ 2 = 300
60 000 ÷ 200 = 300

186

Reflecting

1. Show how to use estimation to check Khaled's calculation of the number of e-mail messages his class receives each year.

2. a) How did renaming 60 000 as 600 hundreds and 200 as 2 hundreds help Khaled with his calculation?
   b) Why is dividing 60 000 by 200 equivalent to dividing 600 by 2?

3. Explain how you can calculate 12 000 ÷ 300 by calculating 120 ÷ 3.

Checking

4. A school has 69 Grade 6 students. Each student receives about four e-mail messages a day.
   a) About how many e-mail messages in one year are likely infected with a virus that affects 1 in every 500 messages?
   b) Use multiplication to check your answer.

Practising

5. Calculate. Use multiplication to check each answer.
   a) 1000 ÷ 50 =
   c) 64 000 ÷ 800 = ■
   b) 50 000 ÷ 200 = ■
   d) 100 000 ÷ 200 = ■

6. A computer store has an order for 40 000 CDs. How many spindles of each number of CDs can they use to complete the order?
   a)
   b)
   c)

   200 CDs — 400 CDs — 500 CDs

7. Archeologists believe Canada's aboriginal peoples have lived in Canada for at least 15 000 years. How many centuries is that?

CHAPTER 6

9 Estimating Quotients

You will need
- a calculator

Goal Use multiplication and rounding to check the reasonableness of a quotient.

Raven's band is ordering bottled water to sell to those attending its 3-day powwow.

The band knows from past powwows that about 750 people will buy bottled water. Each person is expected to drink about 2 bottles per day.

? About how many cases of water should the band order and what will it cost?

$24.95 for 24

Raven's Estimate

I used mental math to calculate the number of bottles we need to order. It's 4500.
To calculate the number of cases, I used my calculator to divide 4500 by 24. I got 187.5 or about 188 cases.

I'll estimate to see if 188 is reasonable.
I multiply to see if 100 is a low or high estimate of the quotient.

$100 \times 24 = 2400$

100 is an underestimate.
I will try an estimate of 200.

A. What answer will Raven get when she tries an estimate of 200? Show your work.

B. How can you use the estimates of 100 and 200 to determine if the calculation of the number of cases needed is reasonable?

C. Calculate the cost of the number of cases needed. Show how to use estimation to check the reasonableness of your calculation.

188

Reflecting

1. How might you use mental math to determine the number of bottles the band needs to order?

2. How would you use mental math to multiply 100 by 24 and 200 by 24?

Checking

3. Another band is ordering bottled water to sell to the 500 people they expect to attend their 4-day powwow. From past sales, they expect to sell about 4 bottles per person each day.
   a) How many cases of water should they order and how much will it cost?
   b) Show how to use estimation to check each calculation.

$14.95 for 12

Practising

4. Is each answer reasonable? Explain.
   a) 3335 ÷ 23 = 145
   b) 47)9305 — 298.979
   c) 9625 ÷ 77 = 225
   d) 6245 ÷ 52 = 120.096
   e) 13)6201 — 477
   f) 48)8220 — 171.25

5. a) Calculate the mean number of assists in a season for each player.
   b) Estimate to show that your calculation is reasonable.

Assists and Seasons by the end of 2004

Player	Assists	Season
Wayne Gretzky	1963	20
Ron Francis	1249	23
Ray Bourque	1169	22
Mark Messier	1193	25

6. The 78 Grade 6 students in Katrina's school hope to raise over $1000 to bring clean water to villages in Indonesia. The students decided to earn money by doing chores.
   a) Calculate the amount each student hopes to raise.
   b) Estimate to show that your calculation is reasonable.

7. Along the side of a 3 km road, there are 40 light posts that are spaced equally.
   a) What is the distance in metres between each pair of light posts?
   b) Estimate to show that your calculation is reasonable.

CHAPTER 6

10 Dividing by Two-Digit Numbers

You will need
- grid paper

Goal Divide a four-digit number by a two-digit number.

The 43 Grade 6 students from Jorge's school received 5160 trees paid for by Tree Canada. On one weekend, they planted 2795 trees. Each student planted the same number of trees.

? How many more trees does each student have to plant the next weekend to plant all the trees?

Jorge's Division

We still have 5160 − 2795 = 2365 trees to plant.

I'll divide 2365 by 43 to calculate the number of trees each student needs to plant.

Step 1 I estimate the number of tens in the quotient by rounding the divisor 43 to 40. I'll try 50, or 5 tens, as part of the quotient.

$40 \times 50 = 2000$ is low
$40 \times 60 = 2400$ is high but very close.

```
         5 0
   4 3 ) 2 3 6 5
         2 1 5 0  ← 4 3 × 5 0 = 2 1 5 0
           2 1 5
```

Step 2 I multiply 43 by 50. I have 215 left to divide.

Step 3 I estimate the number of ones in the quotient by rounding the divisor 43 to 40. I'll try 5, or 5 ones, as the next part of the quotient.

$5 \times 40 = 200$ is low
$6 \times 40 = 240$ is high

```
               5
             5 0
   4 3 ) 2 3 6 5
         2 1 5 0
           2 1 5
           2 1 5  ← 4 3 × 5 = 2 1 5
               0
```

Step 4 I multiply 43 by 5. I have no remainder.

Each student needs to plant 55 trees.

Emilio's Division

I'll use multiplication to divide 2365 by 43.

Step 1 2365 ÷ 43 = ■ means
43 × ■ = 2365.

43 is less than 50, and
I know 40 × 50 = 2000.

So the quotient is greater than 40.

I will multiply 43 by 40.

I have 645 left to divide.

```
4 3 ) 2 3 6 5
      1 7 2 0      4 0
          6 4 5        43 × 40
```

Step 2 645 ÷ 43 = ■ means
43 × ■ = 645.

I know 43 × 10 = 430.

I have 215 left to divide.

```
4 3 ) 2 3 6 5
      1 7 2 0      4 0
          6 4 5
          4 3 0    1 0
            2 1 5      43 × 10
```

Step 3 215 ÷ 43 = ■ means
43 × ■ = 215.

43 × 5 = 215

I have no remainder.

Each student needs to plant 55 trees.

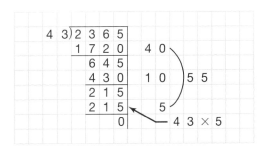

Reflecting

1. Would you use mental math or paper and pencil to calculate 43 × 50 in Jorge's calculation? Justify your choice.

2. Could Emilio have first started by multiplying 43 by 50 instead of multiplying 43 by 40? Explain.

3. Which method of division do you prefer? Justify your choice.

Checking

4. The 58 Grade 5 students in Jorge's school each planted the same number of shrubs. There were 2958 shrubs in total.
   a) How many shrubs did each student plant?
   b) Check your answer by using either multiplication or another division method.

Practising

5. Calculate.
   a) 1441 ÷ 11
   b) 12)1044
   c) 2675 ÷ 27
   d) 75)4545
   e) 7441 ÷ 35
   f) 88)8809

6. a) Estimate to check the reasonableness of one calculation in Question 5.
   b) Use multiplication to check another calculation in Question 5.

7. Use the digits 1, 2, 3, 4, and 5 so the quotient is greater than 200 and the remainder is zero. Use each digit only once.

8. a) Insert a two-digit number. Calculate the result.
      25 × ■ + 13
   b) Divide your answer by 25. What is the remainder?
   c) Try calculating using other two-digit numbers in the box. What do you notice about the remainders?

9. The students in Martha's class collected 1656 cans for recycling.
   a) How many cases of 24 could they fill with these cans?
   b) How many cases of 18 could they fill with these cans?
   c) Estimate to show that each calculation is reasonable.

10. To calculate 3175 ÷ 45, Jeff calculated 3175 ÷ 5 and then divided the quotient by 9. Will this give the correct quotient? Explain.

Math Game

Coin Products

You will need
- a die
- a calculator
- a spinner

Number of players: 2 to 4
How to play: Calculate the value of a money amount.

Step 1: One player spins to get a coin.

Step 2: The player then rolls the die twice to get a two-digit number. The first number rolled represents tens. The second number represents ones. The two-digit number represents the number of the coins obtained in Step 1.

Step 3: The player uses mental math to calculate the value of the coins in cents or dollars. A calculator may be used to check the answer.

? **A correct answer is worth one point. The first player to score 5 points is the winner.**

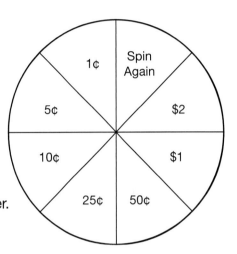

Ayan's Calculation

I spun 25¢.

I rolled 3 and 6. So I have 36 quarters.

Every 4 quarters is worth $1.

So I have $9.00 or 900 cents.

I calculated 36 × 25 on the calculator and got 900.

I score one point.

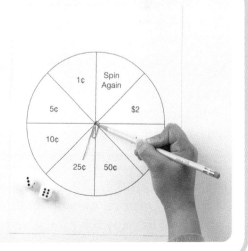

CHAPTER 6

11 Communicate About Creating and Solving Problems

You will need
- a calculator

Goal Create and explain how to solve multiplication and division problems.

Angele's class is using numbers and measurements published in books, magazines, newspapers, and on the Internet to create multiplication and division problems about their lives. The problems and their solutions will be presented in posters at the school's math and science fair.

Angele wanted to use this information to create a problem.

- 1 L of gas burned by a car produces more than 2 kg of carbon dioxide.
- Our van burned 292 L of gas this month.

? How can you explain your solutions to problems that you created?

Angele's Problem and Solution

My Problem
Each litre of gasoline our van burns produces over 2 kg of carbon dioxide. Our van uses 292 L of gas each month. How many tonnes of carbon dioxide does our van produce each year?

My Solution

Understand
I need to determine the number of tonnes of carbon dioxide produced by our van each year.

Make a Plan
I will multiply 292 by 12 to calculate the total number of litres of gasoline our van might use in a year. Then I will multiply the total number of litres by two to get the number of kilograms of carbon dioxide produced by the van.

A. Write up a solution for Carry Out the Plan. Use the Communication Checklist.

Communication Checklist
- ☑ Did you show all the steps?
- ☑ Did you check your answers?
- ☑ Did you show the right amount of detail?

Reflecting

1. Why is it important to check each calculation in each step of a solution to a problem?
2. How could you have improved Angele's explanation?

Checking

3. **a)** Use the newspaper clipping and any other numbers you need to create a multiplication or division problem.
   b) Explain your solution. Use the Communication Checklist.

> On average, a 4-year-old child asks 437 questions a day.

Practising

4. **a)** Choose some of these facts or numbers and measurements collected from magazines, newspapers, the Internet to create a multiplication or division problem.
   b) Explain your solution. Use the Communication Checklist.

> A flea can jump 350 times its body length.

> Each year the Moon moves about 4 cm farther from the Earth.

> The cruise liner Queen Elizabeth II moves only about 4 cm for each litre of diesel that it burns.

195

12 Order of Operations

You will need
- a calculator

Goal Determine whether the value of an expression changes when the order of calculating changes.

Tom entered several contests that offer prizes. To win a prize, he has to answer skill-testing questions. He says there are many possible answers to each question.

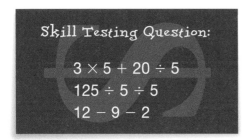

Skill Testing Question:

$3 \times 5 + 20 \div 5$

$125 \div 5 \div 5$

$12 - 9 - 2$

? **What skill-testing questions can you create that have more than one answer?**

A. How many different answers can you get when you answer each skill-testing question? Show your work.

B. For each skill-testing question, enter every number and operation from left to right into your calculator. Then press =. What answer is displayed?

C. Do you get the same answer if you press = each time you enter a number and an operation?

D. Create a skill-testing question with three different whole numbers that contains two identical operations signs. How many different answers can you get?

E. Create a skill-testing question with four different numbers and three different operations. How many different answers can you get?

F. Create a skill-testing question with four numbers that has only one possible answer.

Reflecting

1. Why would calculators be designed to give only one answer when you enter all numbers and all operations before you press ▇ ?
2. Does the order in which you calculate an expression change the answer? Explain using an example.

Curious Math

Egyptian Division

The ancient Egyptians used doubling and halving to multiply and divide numbers.

Suppose you want to divide 153 by 9. $153 \div 9 = \blacksquare$

Write 1 at the top of one column and the divisor 9 at the top of another column. Double each number in both columns until you get a number in the second column that is greater than or equal to 153.

1	9
2	18
4	36
8	72
16	144
32	288 Stop here.

Now identify numbers in the second column that have a sum of 153.

$144 + 9 = 153$

The quotient is the sum of the numbers in the first column opposite the numbers you chose in the second column.

$16 + 1 = 17$

$153 \div 9 = 17$

1. Calculate each quotient. Use Egyptian division.
    a) $125 \div 5$ b) $165 \div 15$ c) $148 \div 4$

CHAPTER 6

Skills Bank

LESSON

1
1. Identify the factors of each number.
   a) 24 b) 49 c) 64 d) 75

2. Blake expressed 36 as these pairs of factors:
   1 × 36, 2 × 18, 3 × 12, 4 × 9, and 6 × 6.
   Use these expressions to identify the factors of 36.

3. a) Was the year 2005 a prime or a composite number?
   b) Name two years after 2005 that will be composite.

2
4. List five multiples of each number.
   a) 4 b) 6 c) 10 d) 11

5. A ringing alarm and a flashing light are turned on at the same time. The bell rings every 2 min. The light flashes every 3 min.
   a) How many times will the bell ring in 30 min?
   b) How many times will the light flash in 30 min?
   c) How many times will the bell ring and the light flash at the same in 30 min.

3
6. Calculate the value of each number of coins.

   a)
   40 quarters

   b)
   50 dimes

7. Explain how you can use money to calculate 12 × 50.

4
8. Calculate each product using mental math.
   a) 50 × 70   b) 40 × 200
   c) 60 × 500 d) 800 × 500

9. What is the area of each rectangle?   a)
   300 cm
   100 cm

   b)
   900 cm
   200 cm

10. Which products are reasonable? Use estimation.
    a) 76 × 89 = 7764
    b) 102 × 48 = 4896
    c) 45 × 4985 = 124 325
    d) 2394 × 63 = 150 822

11. Agnes's family and friends are making 48 vests for their Potlatch. They decided to sew 275 buttons on the traditional design of each vest. A package of 24 buttons costs $2.
    a) Calculate the cost of the buttons.
    b) Show that each of your calculations is reasonable. Use estimation.

12. Calculate.
    a) 54 × 105
    b) 198 × 77
    c) 4306 × 55
    d) 88 × 8888
    e) 3827 × 97
    f) 34 × 904

13. The distance between two bus depots is 378 km. A bus makes two return trips twice each week.
    a) What is the total number of kilometres travelled by the bus each year?
    b) Show that your answer is reasonable.

14. Calculate. Use mental math.
    a) 2000 ÷ 1000
    b) 38 000 ÷ 1000
    c) 457 000 ÷ 1000
    d) 1 000 000 ÷ 1000
    e) 2 770 000 ÷ 1000
    e) 60 000 ÷ 10 000
    f) 650 000 ÷ 10 000
    g) 6 000 000 ÷ 10 000
    h) 6 550 000 ÷ 10 000
    i) 2 770 000 ÷ 10 000

15. Calculate each quotient using mental math. Use multiplication to check each quotient.
    a) 1000 ÷ 20
    b) 200)10 000
    c) 40 000 ÷ 500
    d) 800)480 000

16. Trees remove carbon dioxide from the air. You can estimate the number of tonnes of carbon dioxide removed annually by the trees in a city by dividing the number of trees by 400. If a city has 12 000 trees, about how many kilograms of carbon dioxide will the trees remove from the air each year?

17. Which quotients are reasonable? Use estimation.
    a) 9675 ÷ 25 = 387
    c) 2820 ÷ 12 = 135
    b) 45)9990 with 122 above
    d) 39)7293 with 187 above

18. Explain how you know that 71)4431 has an answer between 50 and 70.

19. A golf club is sponsoring a tournament for 144 golfers. Each golfer will be given 12 free balls. The club has bought golf ball packs containing 36 balls. The packs cost $25.00.
    a) What will it cost the club to buy the balls needed for the tournament?
    b) Show that each calculation is reasonable. Use estimation.

20. Calculate.
    a) 9774 ÷ 18 = ■
    c) 7605 ÷ 65 = ■
    b) 6058 ÷ 13 = ■
    d) 9380 ÷ 35 = ■

21. a) Show how to use estimation to check one quotient in Question 21.
    b) Show how to use multiplication to check another quotient in Question 21.

22. A carton holds 24 eggs. How many cartons do you need to hold 1800 eggs?

23. A bus company expects to transport 2500 people to a football game. How many 48-passenger buses do they need?

24. What is the length of the rectangle?

65 cm | 5525 cm²

■ cm

25. a) Create a multiplication or division problem using the numbers 25 and 1575.
    b) Explain your solution.

26. The winner of a lottery must answer this skill-testing question to collect the prize.
    2 × 4 + 10 × 3 = ■
    The official answer is 38. Explain how this answer was calculated.

CHAPTER 6

Problem Bank

LESSON

1

1. a) Which numbers from 2 to 30 have an odd number of factors?
   b) What do you notice about these numbers?
   c) Predict the next number with an odd number of factors.

2

2. a) List the first five multiples of 99 greater than 99. What pattern do you notice?
   b) Use the pattern to predict the sixth multiple of 99.

3. Some cicadas return every 17 years. Other cicadas return every 13 years. Both kinds appeared together in 2004 in the U.S.
   a) When was the previous year that they both appeared together?
   b) Predict the next year that they will appear together.

4. A pair of numbers has a sum of 100. One number is a multiple of 3. The other number is a multiple of 11. How many pairs can you find?

3

5. Rosa has a collection of 36 coins. The coins are all quarters and half-dollars. The value of the collection is $12. How many of each type of coin does she have? Show your work.

4

6. A micro-sized atomic clock (less than the size of your thumb) has an accuracy that allows it to lose only one second every 300 years. How many years will it take before the clock loses a minute? Show your work.

6

7. a) Calculate the number of days that you have lived.
   b) Use estimation to show that your answer is reasonable.

8. Sound travels about 340 m in a second.
   a) If you see a lightning flash and hear thunder 12 s later, how far away is the lightning?
   b) Use estimation to show that your answer is reasonable.

9. Samuel's grandmother is making Saskatoonberry jam. She uses 500 g of sugar for every 600 mL of berry pulp. She has 30 L of berry pulp. How much sugar will she need to use?

10. Crunch E Bites cereal has distributed prizes among 10 000 cereal boxes. The chart gives information about the prizes. What is the total value of all prizes in the cereal boxes? Show your work.

Prize	Retail Value	Chances
$1 off next box of cereal	$1	1 in 10
$2 off next box of cereal	$2	1 in 20
1 free DVD rental	$5	1 in 40
2 free DVD rentals	$10	1 in 50
Music store gift certificate	$50	1 in 100

11. One side length of a rectangle is between 25 cm and 50 cm. The area of the rectangle is between 8000 cm² and 8500 cm². Use a calculator to solve each problem. Show all your steps.
    a) What is the greatest possible length?
    b) What is the least possible length?
    c) Use estimation to check each answer.

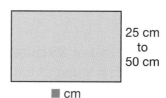

25 cm to 50 cm

■ cm

12. Four different four-digit numbers are divided by 13. The remainder each time is 10. What might the four numbers be?

13. A shape with a perimeter of 1870 cm is made from 10 squares.
    a) What is the side length of each square?
    b) What is the total area of the shape?

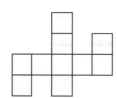

CHAPTER 6

Frequently Asked Questions

Q: How do you divide by tens and hundreds?

A: You can divide by renaming each number to get the same units before dividing.

For example, you can divide 120 000 by 200 using these steps:

120 000 ÷ 200 = ■
1200 hundreds ÷ 2 hundreds = ■
1200 ÷ 2 = 600
Check: 200 × 600 = 120 000

Q: How can you estimate quotients?

A: You can estimate a quotient by rounding the divisor up or down to a number of tens. Then you can multiply by hundreds to get an underestimate and an overestimate.

For example, $24\overline{)6456}$ gives 269.

$24\overline{)6456}$ is between $20\overline{)6456}$ and $30\overline{)6456}$

$$\begin{array}{cc} 300 & 200 \\ 20\overline{)6456} & 30\overline{)6456} \\ 6000 & 6000 \end{array}$$

The quotient is between 200 and 300. So 269 seems reasonable.

Q: How can you use multiplication and addition to check answers when dividing?

A: You can check an answer when you divide by multiplying the divisor by the quotient and adding any remainder.

For example, $56\overline{)4017}$ gives $71\ R41$.

56 × 71 = 3976
3976 + 41 = 4017

CHAPTER 6

Chapter Review

LESSON

1
1. Identify the factors of each number.
   a) 12   b) 17   c) 26   d) 60

2. Identify each number in Question 1 as prime or composite. Explain.

2
3. Identify three multiples of each number.
   a) 7   b) 9   c) 75   d) 300

3
4. Use the relationship between coin values to calculate 48 × 5.

4
5. Use mental math to calculate each product.
   a) 50 × 80
   b) 200 × 100
   c) 50 × 200
   d) 900 × 900

6
6. Calculate. Estimate to show that each answer is reasonable.
   a) 25 × 58
   b) 5 × 3333
   c) 55 × 552
   d) 8785 × 67
   e) 93 × 8576
   f) 1782 × 25

7. A carton contains 12 eggs. Egg cartons are packed into crates. Each crate contains 12 cartons.
   a) How many eggs are in 25 crates?
   b) Estimate to show that each calculation is reasonable.

7
8. Calculate. Use mental math.
   a) 26 000 ÷ 1000
   b) 630 000 ÷ 1000
   c) 63 000 ÷ 10 000
   d) 1 650 000 ÷ 10 000

8
9. Calculate each quotient using mental math. Use multiplication to check each quotient.
   a) 15 000 ÷ 10
   b) 500)‾100 000
   c) 40 000 ÷ 800
   d) 700)‾490 000

10. A rancher is building a wire fence to keep her bison inside her ranch. She needs wire to surround a 500 m by 800 m field.
    a) How much will it cost to use each length of wire to make the fence?
    b) How much money will she save by buying the best bargain instead of the worst bargain?

100 m for $180   50 m for $100   20 m for $50

11. Estimate each quotient.
    a) 4403 ÷ 27 = ■
    b) 3125 ÷ 25 = ■
    c) 8570 ÷ 55 = ■
    d) 6993 ÷ 37 = ■

12. Calculate each quotient in Question 11. How did your estimate compare to the calculated value?

13. For every 25 mm of rain falling on the average home, 5000 L runs off the roof. Some people use barrels to collect this rain water, which is used for watering plants. Cecilia's town usually has 2625 mm of rain each year.
    a) How many litres of water can be collected by using rain barrels in Cecilia's town? Show your work.
    b) Estimate to show that each calculation is reasonable.

14. Create and solve a multiplication or division problem using the information.
    a) A section in a stadium has 23 rows of 175 seats.
    b) A rectangle has a width of 50 cm and an area of 2400 cm².

15. a) Show how to get an answer of 50 when calculating 10 × 3 + 4 × 5.
    b) List another possible answer.

CHAPTER 6

Chapter Task

Chartering Buses

Imagine your school needs to charter buses to take the entire school on a day-long field trip.

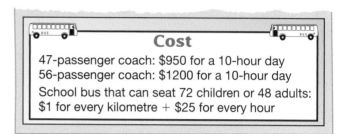

Cost
47-passenger coach: $950 for a 10-hour day
56-passenger coach: $1200 for a 10-hour day
School bus that can seat 72 children or 48 adults:
$1 for every kilometre + $25 for every hour

? **What is the cost of chartering the buses?**

A. Decide on a trip to a place that you would like to visit. Determine the total distance for a return trip.

B. Determine the number of students and adults who will go on the trip.

C. Calculate the number of each type of bus you will need to charter. Show your work.

D. What will it cost to charter each type of bus for your trip? Show your work.

E. What will you charge each student and adult to pay for the buses? Show your work.

Task Checklist
☑ Did you estimate to check that each calculation is reasonable?
☑ Did you explain how you calculated?
☑ Did you show all your steps?
☑ Did you explain your thinking?

CHAPTER 7

2-D Geometry

Goals

You will be able to
- describe properties of triangles and quadrilaterals
- construct polygons using measurement tools
- sort polygons according to line symmetry and other properties
- communicate about polygons

Flying kites

CHAPTER 7

Getting Started

You will need
- a transparent mirror
- a protractor
- a ruler

Mystery Design

One of these polygons contains Chandra's mystery design.

A.

C.
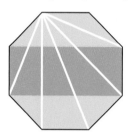

B.

D.
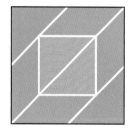

She made clue cards to help identify the mystery design.
The yellow cards give clues about the polygon.
The red cards give clues about the design on the polygon.

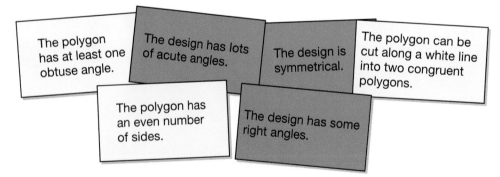

? **How can you identify the mystery design?**

A. Describe each polygon. Which ones have the right number of sides to be the mystery polygon?

B. Are any other polygons eliminated by the yellow cards? Which ones? Why?

C. Are any other polygons eliminated by the red cards? Which ones? Why?

D. Which polygon contains Chandra's mystery design?

Do You Remember?

1. Classify the following triangles two ways: first by angle measures and then by side lengths. Use the words obtuse, right, acute, equilateral, isosceles, and scalene.

   a)

   c)

   b)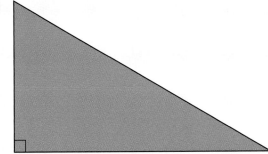

   d)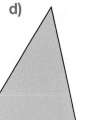

2. Construct a triangle with a 73° angle between two sides with lengths of 3 cm and 5 cm.

3. Identify each angle in this polygon as acute, obtuse, right, or straight.

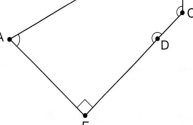

CHAPTER 7

Estimating Angle Measures

You will need
- a protractor
- a ruler
- an air hockey table model

Goal Compare and estimate angle measures.

Denise is preparing for an air hockey tournament. She is working on shots on net that bounce off the side board once.

? At about what angle should a puck bounce off the side board to result in a shot on net?

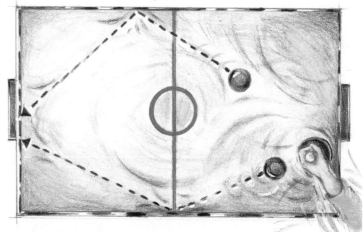

Denise's Sketches

I drew two sketches to show the path of the puck.

In each sketch, I show a dot where the puck starts and lines to represent the path of the puck.

I labelled the angles between the path of the puck and the side boards.

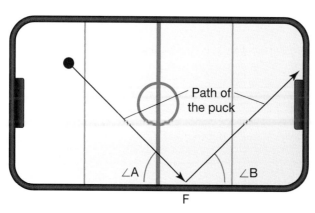

210

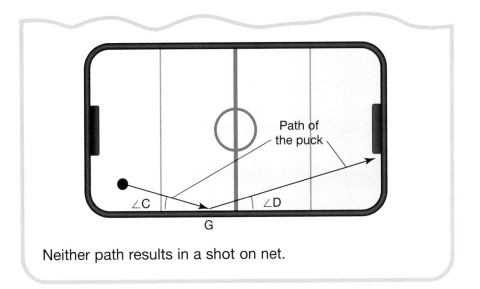

Neither path results in a shot on net.

A. Denise says ∠A is about 45°. Do you agree? Explain. Test Denise's estimate by measuring.

B. Estimate ∠B, ∠C, and ∠D. Test your estimates. What do you notice about the pairs of ∠A and ∠B, ∠C and ∠D?

C. If Denise bounces the puck off point F, about how much should ∠B decrease to result in a shot on net?

D. If Denise bounces the puck off point G, about how much should ∠D increase to result in a shot on net?

E. Choose a different point on the side board. Estimate an ∠E and ∠F that will result in a shot on net. Test your estimate.

Communication Tip
Angles are noted using the "∠" symbol. We read ∠A as "angle A."

Reflecting

1. What strategies did you use to estimate the measures of your angles?

2. a) What do you notice about the path of the puck if ∠E decreases?
   b) What do you notice about the path of the puck if ∠E increases?

3. What will happen if the puck hits the side board at a right angle?

CHAPTER 7

2 Investigating Properties of Triangles

You will need
- a ruler
- a protractor

Goal Investigate angle and side relationships of triangles.

Khaled's relay team is going to compete on a triangular course. Each runner runs one side length of the triangle. Khaled has one slow runner on his team. He thinks his team can win if the course has a short side for his slow runner.

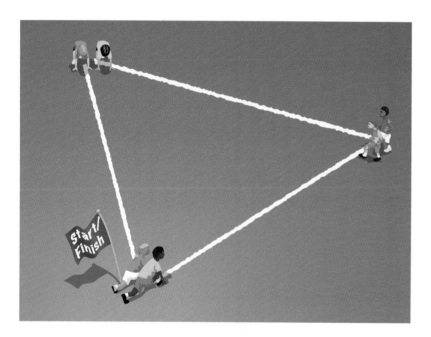

Course Rules
- total distance between 200 m and 250 m
- angles between sides are at least 40°
- each side is at least $\frac{1}{4}$ of the total

? **How can you design a course so that Khaled's team is more likely to win?**

A. Draw some possible triangular courses. Include some that are symmetrical and some that are not. For each course, make sure you follow the course rules.

B. Label each triangle so that the angles are labelled A, B, and C and the sides **opposite** them are labelled a, b, and c, in that order.

C. Measure the angles and side lengths for each triangle you draw.

D. Order the side lengths of each triangle from least to greatest.

opposite side
In a triangle, the side an angle does not touch.
For example, the opposite side of ∠Q is q.

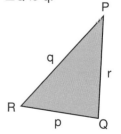

E. Order the angle sizes of each triangle from least to greatest.

F. Which of the courses you've drawn will best help Khaled's team win?

Reflecting

1. How did you make sure that your courses met the requirements?
2. What strategy did you use to get the best course?
3. Compare the side length and angle size orders in Parts D and E. What do you notice?

Mental Imagery

Visualizing Symmetrical Shapes

You will need
- a transparent mirror

The coloured part of each drawing is $\frac{1}{4}$ of a shape that has a vertical and a horizontal line of symmetry.

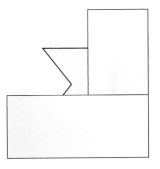

A. Visualize the whole shape.
B. Draw your visual image of the shape.
C. Check that your drawing has both lines of symmetry.

Try These

1. Complete each shape.

a) 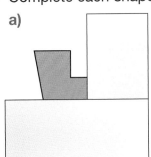   b) 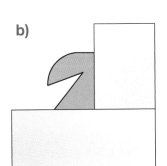   c)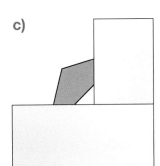

CHAPTER 7

3 Communicate About Triangles

Goal Communicate and explain geometric ideas.

Emilio measured some triangles and noticed something true about all of them. He made a **hypothesis**.

? How can Emilio explain his hypothesis?

hypothesis
A statement that you think you can test

Communication Tip
The plural of hypothesis is hypotheses.

Emilio's Hypothesis

When I add the lengths of two sides of a triangle, the sum is always greater than the other length.
I think this is always true.

I'll test my hypothesis with more examples.

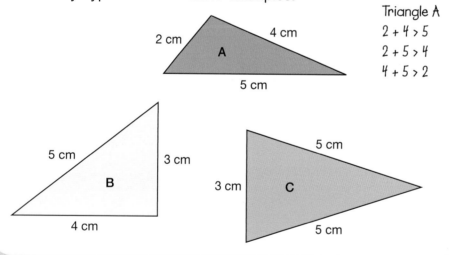

Triangle A
2 + 4 > 5
2 + 5 > 4
4 + 5 > 2

A. Did Emilio state his hypothesis clearly?
B. Did Emilio communicate his hypothesis and his reasons using mathematical language?
C. Was it clear how Emilio would choose examples to test his hypothesis?
D. How can you improve Emilio's communication?

Communication Checklist
☑ Did you use math language?
☑ Did you explain your thinking?
☑ Did you include diagrams?

Reflecting

1. Why is it important to explain your hypothesis clearly?

Checking

2. Raven noticed that in an isosceles triangle, there seem to be two equal angles as well as two equal sides.
   a) What relationship do you notice between the equal angles and the equal sides in each isosceles triangle shown?
   b) Draw scalene triangles. Are any of the angles equal?
   c) Write your hypothesis. Explain how you could test it. Use the Communication Checklist.

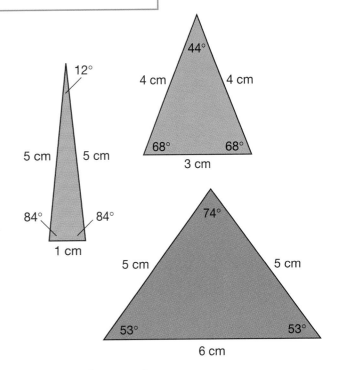

Practising

3. a) What do you notice about the number of lines of symmetry in equilateral, isosceles, and scalene triangles?
   b) Make a hypothesis about the number of lines of symmetry for these triangles based on your observation.
   c) Test your hypothesis.
   d) Exchange your hypothesis with a classmate. How can you improve on your partner's communication? Use the Communication Checklist.

4. a) Calculate the sum of the two shorter sides of the triangle. Double the sum. Compare the doubled sum to the perimeter of each triangle. Make a hypothesis about this relationship.
   b) Explain how you will test your hypothesis.

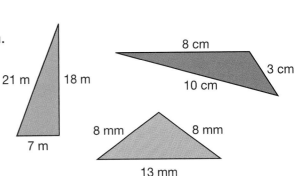

CHAPTER 7

Frequently Asked Questions

Q: How can you estimate the size of an angle?

A: You can compare it to some angles you know.

For example, ∠A is a little less than 45°, so it is probably about 40°. ∠B is between 60° and 90°, but closer to 60°. It's probably about 70°.

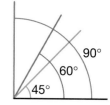

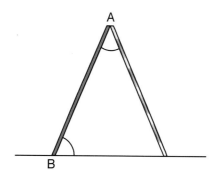

Q: In a triangle, how are angle measures related to side lengths?

A: In a triangle, the largest angle is opposite the longest side, and the smallest angle is opposite the shortest side. In triangles where angles are equal, the sides opposite the equal angles are also equal.

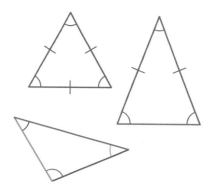

For example, the lengths of the sides of this triangle are 1.0 m, 1.6 m, and 1.7 m. How can you label them without measuring? The smallest angle is 35°, so label the opposite side 1.0 m. The largest angle is 78°, so label the opposite side 1.7 m. Label the remaining side 1.6 m.

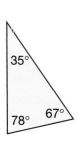

CHAPTER 7
Mid-Chapter Review

1. **a)** Estimate the sizes of ∠A, ∠B, and ∠C. Explain your strategy.
   b) Measure to check your estimates. How close were your estimates?

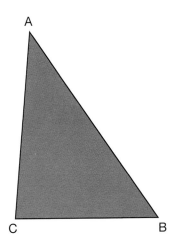

2. Trace the triangle. Without using a protractor or a ruler, label the side lengths and angle measures with these measurements.

20°	60°	100°
5.9 m	15 m	17 m

3. Irene was in a hurry when she measured some triangles and recorded these measurements. Identify possible errors and suggest what she may have measured incorrectly. Explain how you know.
   a) 60°, 60°, 60° and 5 cm, 5 cm, 6 cm
   b) 80°, 80°, 20° and 5 cm, 5 cm, 1.7 cm
   c) 80°, 80°, 20° and 4 cm, 5 cm, 6 cm

4. Trade your answers for Question 3 with a partner. How could you improve on your partner's explanations?

CHAPTER 7

4 Constructing Polygons

You will need
- a ruler
- a protractor

Goal Construct polygons based on angle measures and side lengths.

Communication Tip
When a shape is constructed, it is drawn carefully with appropriate tools, such as rulers and compasses, to meet the requirements of the task. "Draw" often means "construct."

The students made measurements for props to be built for a school play. Rebecca's job is to construct accurate scale diagrams of the props for the builder.

? How can you draw the scale diagrams?

Rebecca's Diagram

The set designer gave me this incomplete sketch of a parallelogram.

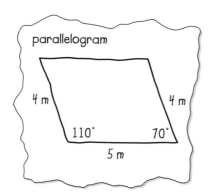

My paper is not big enough for me to draw a diagram with side lengths in metres.

I'll draw with a scale of 1 cm represents 1 m. Then I can draw in centimetres instead of metres.

I start by drawing and labelling the 5 cm side.

Then I construct the 110° angle at one end of the side. 110° is 20° past the 90° angle mark.

218

I draw the 4 cm side.

At the other end of the 5 cm side, I construct the 70° angle and another side of 4 cm.

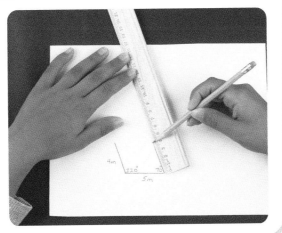

A. Draw and finish Rebecca's parallelogram.

B. Measure the angles that were not given in the sketch.

C. Draw a scale diagram of this **kite**. Use a scale of 1 cm = 1 m.

kite
A quadrilateral that has two pairs of equal sides with no sides parallel

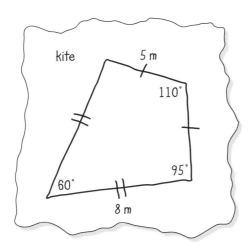

Reflecting

1. In what order would you draw the sides and angles of a parallelogram?

2. In what order did you draw the sides and angles of the kite in Part C? Why does this order make sense?

Checking

3. Draw this polygon. Use a scale of 1 cm represents 1 m. Measure and label the unlabelled sides and angle.

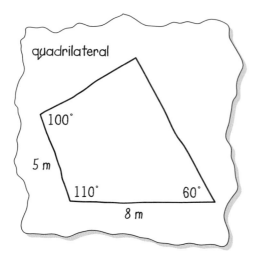

Practising

4. Draw these polygons.

   a) parallelogram

   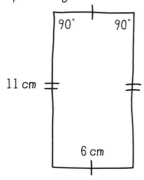

   b) parallelogram

   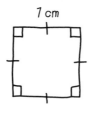

   c) regular pentagon

   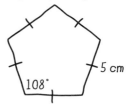

5. Draw a rhombus with side lengths of 10 cm and angle measures of 40° and 140°.

6. The measurements of the sides and angles in this parallelogram are all correct, but the set designer put some of the labels in the wrong places.
   a) What possible errors were made in this sketch? Explain.
   b) Draw a scale diagram of the parallelogram.

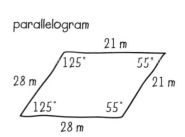

Curious Math

Folding Along Diagonals

You will need
- quadrilateral shapes
- a ruler
- a protractor

Tom's folding

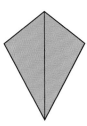

I notice that if I fold along this diagonal line I have two congruent triangles as halves.

But if I fold along this diagonal line then I get two different triangles.

1. Draw and cut out a different quadrilateral. Fold it along its diagonals. Determine if the halves are congruent triangles.

2. Repeat with other quadrilaterals. Record your results.

3. What are the characteristics of quadrilaterals that give congruent triangles when diagonals are folded?

CHAPTER 7

5 Sorting Polygons

You will need
- a ruler
- scissors
- polygon shapes
- a transparent mirror

Goal Sort polygons by line symmetry.

Kurt has a kite collection. He says that $\frac{1}{3}$ of his kites have exactly four lines of symmetry.

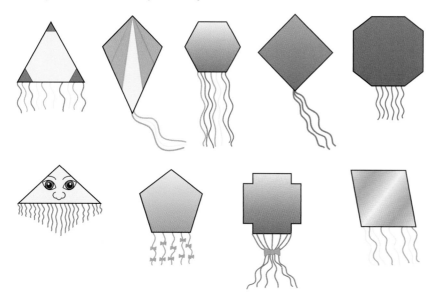

❓ **Which of the kites have four lines of symmetry?**

Ayan's Sorting

I'll trace and cut out these polygons.

I'll use a mirror to determine the lines of symmetry.

Then I will check by folding on the lines of symmetry.

I'll sort the polygons by the number of lines of symmetry.

I'll put the ones with four lines of symmetry together.

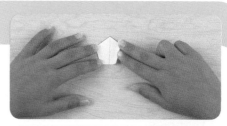

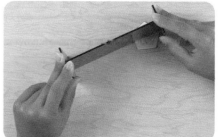

A. Trace and cut out Kurt's kites.

B. Identify and label the lines of symmetry. Use a mirror.

C. Record the number of lines of symmetry of each polygon.

Reflecting

1. What clues did you use to help you locate shapes with four lines of symmetry?

2. Were there any polygons with more sides than lines of symmetry? Were there any polygons with more lines of symmetry than sides?

3. For regular polygons, how can you quickly predict the numbers of lines of symmetry?

Checking

4. Sort Kurt's kites according to number of lines of symmetry and number of equal sides. Use a Venn diagram.

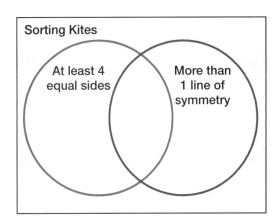

Practising

5. a) Identify the lines of symmetry of each of these polygons.
   b) Sort the polygons. Use the Venn diagram from Question 4.

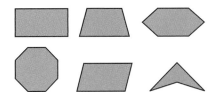

6. Predict the number of lines of symmetry for a regular decagon (a polygon with 10 sides). Show your prediction in a sketch.

CHAPTER 7

6 Investigating Properties of Quadrilaterals

You will need
- a protractor
- a ruler
- straws

Goal Sort and classify quadrilaterals by their properties.

Qi created a poster about quadrilaterals for the bulletin board.

As he drew them, he noticed that some had diagonals that intersected at 90° angles and some didn't. He also noticed that some had diagonals that were equal and some didn't.

He made a hypothesis.

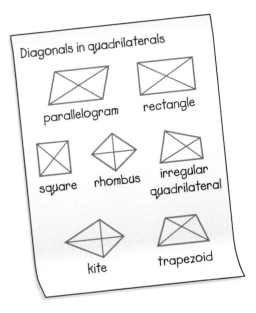

Squares have
- equal diagonals
- diagonals that meet at 90°

? How can you test Qi's hypothesis?

A. Construct quadrilaterals with unequal diagonals to test Qi's hypothesis. Use two straws of different lengths as the diagonals. Use the ends of the straws as vertices. Draw each quadrilateral. Include the diagonals in your drawing.

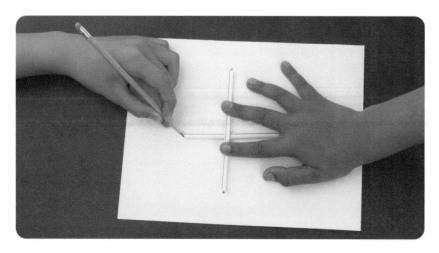

B. Describe each quadrilateral. Are any of your quadrilaterals squares?

C. Construct quadrilaterals with equal diagonals to continue to test Qi's hypothesis. Use two straws of equal length as the diagonals. Use the ends of the straws as vertices.
Draw each quadrilateral. Include the diagonals in your drawing. Measure the angles where the diagonals meet.

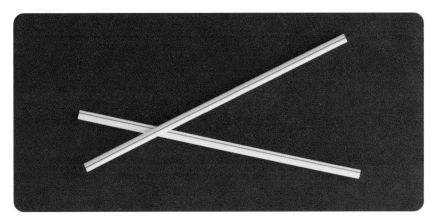

D. Describe each quadrilateral. Are any of your quadrilaterals squares?

E. Is every quadrilateral that has equal diagonals and diagonals that meet at 90° a square? Explain using an example.

F. How can you improve Qi's hypothesis?

Reflecting

1. What do you notice about the angles created where the diagonals meet in all of the quadrilaterals you constructed?

2. a) What kinds of quadrilaterals were you able to make in Part A?
   b) What kinds of quadrilaterals were you able to make in Part C?

3. Make a hypothesis to describe the diagonals of each kind of quadrilateral.

CHAPTER 7

Skills Bank

LESSON

1
1. a) Estimate the sizes of ∠A and ∠B.
   b) Explain what angles you used to help you estimate. Give reasons for your choice of angles.
   c) Measure and check your estimates for ∠A and ∠B.

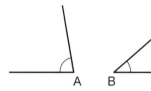

2
2. The triangle's angles measure 110°, 20°, and 50°. The sides have lengths 4 cm, 9 cm, and 11 cm.
   a) Trace the triangle.
   b) Label the angles and side lengths without using a protractor or a ruler. Explain your reasoning.

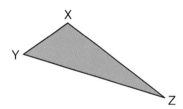

3. Isaac's measurements for the angles for these triangles are: 60°, 35°, 60°, 90°, 55°, and 60°.
   a) Trace the triangles.
   b) Label the angle measures without using a protractor. Explain your reasoning.

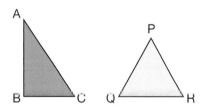

4
4. Construct a regular hexagon with side lengths of 8 cm and angle measures of 120°.

5. Construct a parallelogram with the following measures: angles: 120° and 60°, sides: 4 cm and 6 cm.

6. Trace and cut out these polygons.
   a) Identify and label the lines of symmetry.
   b) Sort the polygons according to the number of lines of symmetry.

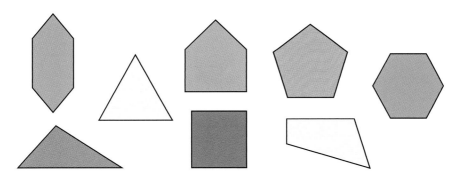

7. A regular polygon has eight lines of symmetry.
   a) Sketch the polygon.
   b) Draw the lines of symmetry.
   c) Was the polygon you sketched the only possible polygon? Explain.

8. What properties do the diagonals of these quadrilaterals have in common?

   a)

   b)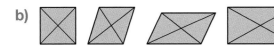

   c)

CHAPTER 7

Problem Bank

LESSON

1
1. A puck bounces off the side of the air hockey game as shown. Selina estimates both the blue angles to be 50°. How can she know, without measuring, that her estimate is close?

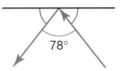

4
2. a) Construct 3 different quadrilaterals where three of the side lengths are 3 cm, 6 cm, and 8 cm.
   b) Were any of the shapes rectangles? Explain.
   c) What's the longest fourth side you made?
   d) Could you have made one with a fourth side of 20 cm? Explain.

5
3. A polygon has been folded in half twice to make this rectangle.
   a) What could the original polygon look like?
   b) Is there more than one answer? Explain.

6
4. Which polygons could the clues refer to? Explain your answers.
   a) My number of lines of symmetry is equal to the number of sides. Who am I?
   b) I have two pairs of parallel sides. My diagonals are perpendicular, but they are not equal. Who am I?
   c) I have two pairs of parallel sides. I have two obtuse angles. My diagonals are not perpendicular. Who am I?

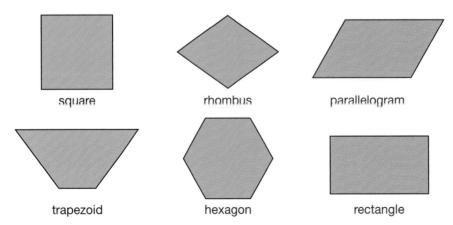

CHAPTER 7

Frequently Asked Questions

Q: How can you construct a parallelogram when you know its angle sizes and side lengths?

A: Start by constructing one side. For example, if the sides of the parallelogram are 7 cm and 4 cm, draw the 7 cm side.

7 cm

Then draw the angles on either end of it. For example, if the parallelogram has 110° and 70° angles, then measure those angles. Make the two new sides the right length.

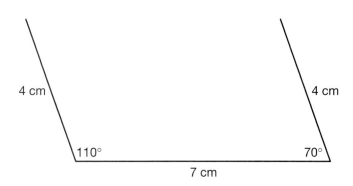

Draw the last side. Check that the last angle is correct by measuring. Label the measures of all the sides and angles.

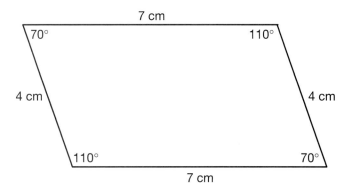

Q: How are the number of lines of symmetry of a regular polygon related to the number of sides?

A: The number of lines of symmetry of a regular polygon is the same as the number of sides.

Equilateral triangles have three sides and three lines of symmetry.
Squares have four sides and four lines of symmetry.
Regular pentagons have five sides and five lines of symmetry.
Regular hexagons have six sides and six lines of symmetry.

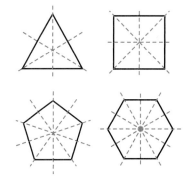

CHAPTER 7
Chapter Review

LESSON 1

1. a) Trace triangle ABC. Estimate its angles without using a protractor.
   b) Which angles could you have used to estimate these angles? Explain why these angles are good choices.

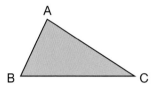

LESSON 2

2. The side lengths of triangle ABC in Question 1 are 16 cm, 27 cm, and 30 cm.
   a) Label the side lengths without using a ruler.
   b) Explain how you know how to label the side lengths.

3. a) The side lengths of this right triangle are 3 cm, 4 cm, and 5 cm. How would you label the side lengths without a ruler? Explain your strategy.
   b) Measure to check.

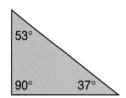

LESSON 3

4. a) Measure the angles in these right triangles.
   b) Make a hypothesis about the value of the sum of the two smaller angles. Use mathematical language.

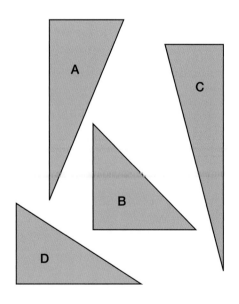

230 NEL

5. The side lengths of a parallelogram are all either 6 cm or 3 cm. Two angles measure 110° and 70°. Construct a parallelogram that matches this description.

6. a) Construct a scalene triangle that has an obtuse angle.
   b) Construct a quadrilateral with a 50° angle, a 95° angle, and a side that is 4 cm long.
   c) Construct a regular pentagon with side length of 4 cm and angle measure of 108°.

7. Order these polygons from least to most lines of symmetry.

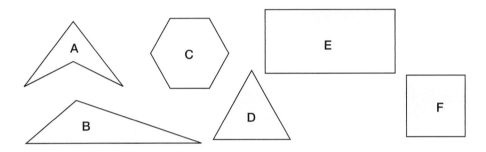

8. Sketch at least three polygons for each description. If a polygon is not possible, explain why.
   a) a regular polygon with at least four lines of symmetry
   b) a regular polygon with six lines of symmetry
   c) a polygon with fewer than four lines of symmetry

9. Sort these quadrilaterals using a Venn diagram. Choose categories from the Properties list.

Properties
- equal diagonals
- perpendicular diagonals
- diagonals all inside the shape

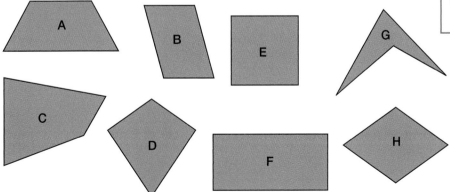

CHAPTER 7

Chapter Task

Furnishing a Bedroom

This is a bird's-eye view of a bedroom.
As the designer, you will choose furniture for the room.

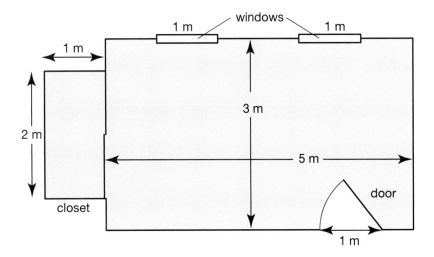

Guidelines for furnishing the room:
- Include at least three shapes with more than one line of symmetry.
- Include a shape with more than five lines of symmetry.
- Include shapes that contain angles of about 20°, 40°, and 120°.
- Include a shape with diagonals that meet at a right angle.
- Include at least three shapes with diagonals that are not equal in length.

? How can you furnish the bedroom?

Represent the top view of the furniture. Use polygons.

A. Draw the bedroom. Use a scale of 4 cm = 1 m.

B. Mark the door, windows, and closets.

C. Construct polygons to represent furniture whose top views match the guidelines.

D. Draw or paste the polygons on the diagram of the bedroom.

E. Explain how your selections meet the guidelines by labelling each polygon with its properties.

Task Checklist
- ✓ Did you include a labelled diagram?
- ✓ Are your explanations clear and complete?
- ✓ Did you use math language?

CHAPTERS 4–7

Cumulative Review

Cross-Strand Multiple Choice

1. The highest mountain on Earth is Mauna Kea, Hawaii. Its height from the floor of the ocean is 10 203 m, but 6033 m of it is below the ocean. About which height is above the ocean?
   A. 16 236 m   B. 4000 m   C. 3000 m   D. 16 000 m

2. In St. John's, Newfoundland and Labrador, the precipitation is generally 0.135 m in January, 0.125 m in February, and 0.117 m in March. Which is the total precipitation for these months?
   A. 0.377 m   B. 0.477 m   C. 0.367 m   D. 0.387 m

3. In the 2004 Olympic Games, Emilie Lepennec of France placed first in the final women's uneven bars with a score of 9.687. Svetlana Khorkina of Russia placed eighth with a score of 8.925. Which is the difference between the scores?
   A. 1.762   B. 18.612   C. 0.762   D. 0.562

4. Which unit is most appropriate to measure the thickness of a string?
   A. millimetres   C. metres
   B. centimetres   D. kilometres

5. Which would you measure in metres?
   A. the distance between cities   C. the height of your desk
   B. the perimeter of the gym   D. the length of a pencil

6. The largest recorded snail is an African giant snail that measured 0.393 m in length. Which length equals 0.393 m?
   A. 393 cm   B. 39.3 mm   C. 3.93 cm   D. 393 mm

7. Angele drew a parallelogram with a perimeter of 18.4 cm. The length of one side is 4.5 cm. Which is the length of another side?
   A. 13.9 cm   B. 9.4 cm   C. 4.7 cm   D. 4.6 cm

8. Which number is a prime number?
   A. 61   B. 42   C. 21   D. 51

9. The distance between Ottawa and Edmonton is 3574 km. Jacob drove this distance 54 times last year. Which is the total distance Jacob drove between these cities?
   A. 32 166 km   C. 182 996 km
   B. 321 660 km   D. 192 996 km

10. 63 students are sharing the cost of a school trip equally. The total amount is $5607. Which is the cost for each student?
    A. $9   B. $809   C. $89   D. $80

11. Kurt drew a triangle with side AB 6.2 cm, BC 3.1 cm, and AC 4.5 cm. Which is true for the triangle?
    A. The triangle has one line of symmetry.
    B. The angle opposite AB is the largest angle.
    C. The triangle is isosceles.
    D. The angle opposite BC is the largest angle.

12. Which is the measure of the largest angle in this quadrilateral?
    A. 125°   C. 130°
    B. 100°   D. 115°

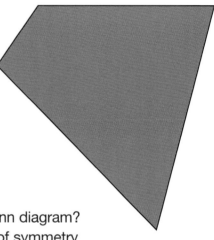

13. In which place does this polygon belong in the Venn diagram?
    A. only the circle labelled: More than three lines of symmetry
    B. only the circle labelled: At least three equal sides
    C. the overlapping section of the circles
    D. outside both circles

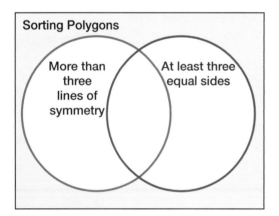

14. Which is not true about a rhombus?
    A. It can have two acute angles and two obtuse angles.
    B. The diagonals are always equal.
    C. It always has four equal sides.
    D. The diagonals always meet at 90°.

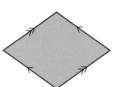

Cross-Strand Investigation

Farming

Denise and Khaled are planning a booth about farming in Ontario for a fall fair.

15. a) Denise read that in 2001, there were 59 728 farms in Ontario, 53 652 in Alberta, and 21 836 in British Columbia. About how many farms were in these three provinces altogether? Explain your reasoning.
    b) About how many more farms were in Ontario than British Columbia in 2001? Explain your reasoning.
    c) Ploughing competitions are organized by the Ontario Plowman's Association. A requirement for the match is that each furrow must be at least a specific width. What unit is appropriate for measuring the width of a furrow? Explain.
    d) What unit is appropriate for measuring the distance a tractor is driven to plough four large fields? Explain.
    e) Khaled found out that in 2002, Ontario farms produced 810 t of fresh plums, and 26 times that amount of peaches, pears, and plums altogether. What was the total mass for peaches, pears, and plums?
    f) Denise measured two quadrilateral fields. In each field, the opposite sides are equal. The diagonals of the corn field are equal, but they do not intersect at 90°. The diagonals of the hay field are not equal, and do not intersect at 90°. What kind of quadrilateral is each field? Sketch diagrams to show your answer.

16. a) Denise did research on the number of farms in Ontario. She recorded the data in this chart. Round the number of farms to the nearest thousand. Create a graph with the vertical axis labelled Thousands of farms. Justify your choice of graph.
    b) Khaled found data about farm operators on the Internet. He recorded the data in this chart. Write a fraction and a decimal to describe each age group. Write each decimal in expanded form.

Farms in Ontario

Year	Number of farms
1981	82 448
1986	72 713
1991	68 633
1996	67 520
2001	59 728

Ontario Farm Operators by Age for Every 1000 in 2001

Age group	Number for every 1000
Under 35 years	106
35–54 years	519
55 years and over	375

Cross-Strand Investigation

4-H Programs

Chandra and Rodrigo researched information about the 4-H program for the fall fair. 4-H stands for Head, Heart, Hand, and Health. It is mainly a program for youths in rural areas, but people in urban areas also join.

17. **a)** Rodrigo met members of a 4-H club who visited a maple syrup farm and brought home 10 L of maple syrup. They used 2.650 L of maple syrup for a squash and carrots recipe and 1.585 L for muffins. How much maple syrup do they have left?

b) Chandra met members of a 4-H club who are making a quilt. Draw a rectangle patch and a scalene triangle patch, each with a perimeter of 16.4 cm. Label the length of each side.

c) Draw a rhombus quilt patch with side lengths of 5 cm and angle measures of 45° and 135°. Draw the lines of symmetry.

d) Chandra talked to members of a 4-H club who are growing pumpkins. The masses of their largest pumpkins were 5.803 kg, 3.275 kg, and 4.006 kg. What is the total mass of the three pumpkins?

e) Each of the 22 members of a 4-H club decided to e-mail 187 people from Alberta they met at a conference. How many e-mails did they send?

f) What unit is appropriate for measuring the distance between Alberta and a 4-H club house in Ontario?

g) Chandra researched these data about 4-H clubs. Chandra says that the mean number of projects completed by a club is about 13. Do you agree? Justify your answer.

Ontario 4 H Clubs in 2004

Members	6619
Clubs	934
Projects	12 412

18. **a)** In Canada, 4-H started in Roland, Manitoba in 1913. Write a pattern rule for the 10th, 20th, 30th, … anniversaries. What are the 1st term and the common difference? What year is the 100th anniversary? Use your pattern rule.

b) Create a survey question about projects that students in your class might like to do in a 4-H club.

c) The 4-H Ontario Magazine is mailed three times a year. In a recent year, 109 648 magazines were mailed altogether. Write the number in expanded form in two ways.

CHAPTER 8

Area

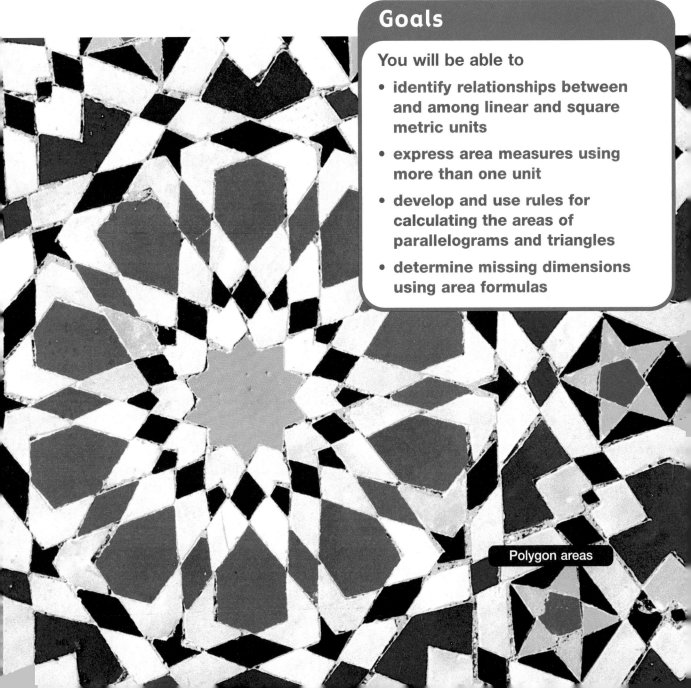

Goals

You will be able to

- identify relationships between and among linear and square metric units
- express area measures using more than one unit
- develop and use rules for calculating the areas of parallelograms and triangles
- determine missing dimensions using area formulas

Polygon areas

CHAPTER 8

Getting Started

You will need
- pencil crayons
- 1 cm grid paper transparency
- 1 cm grid paper
- a ruler

Area Puzzle

Marc made an area challenge by making a turtle puzzle from geometric shapes.

? **How can you design a puzzle from geometric shapes that can be used as an area challenge?**

A. Determine the area of each different colour used in Marc's turtle. Show your work.

B. Design your own puzzle on grid paper. Include a triangle, a rectangle, a square, a shape with a curved edge, and a shape made from more than one rectangle. Use at least four colours in your puzzle.

C. Challenge a partner to determine the area of each colour used in your puzzle.

Do You Remember?

1. Estimate and then measure the area of each polygon in square units. How close were your estimates?

   a)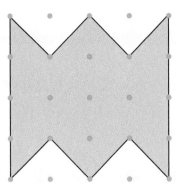
   b)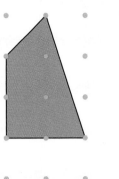

2. Calculate the area of each rectangle. Use the rule for area of a rectangle. Show your work.

   a)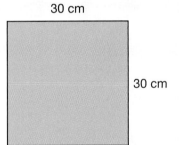
   b)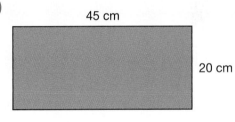

3. Would you measure each area in square kilometres, square metres, square centimetres, or square millimetres?
   a) provincial park
   b) glass for a poster frame
   c) table cloth for a picnic table
   d) calculator screen

4. The community centre is adding a skateboard park. It will be a rectangle with dimensions of 45 m by 60 m. A model will be used to decide where the ramps will be positioned. The model has a scale of 1 cm represents 3 m.
   a) What is the area of the skateboard park?
   b) What is the area of the model?
   c) If a region of 81 m² is roped off as a waiting area, what will be the area of the roped-off region on your model?

CHAPTER 8

1 Unit Relationships

You will need
- base ten blocks
- a calculator

Goal
Identify relationships between and among linear and square metric units.

A tile store has donated 90 000 tiles that are 1 cm² each for a community pool renovation. Rodrigo's class has designed three mosaics that they will make from the tiles to decorate the entrance to the pool.

1 m²

2 m²

4 m²

← door →

? Will Rodrigo's class have enough tiles for all three mosaics?

Rodrigo's Solution

I will start by determining the number of 1 cm² tiles needed for the first mosaic. It is a square with an area of 1 m².

I know the area of a rectangle is the product of the length and the width.

I will use rods as 10 cm rulers to model the **dimensions** of the first mosaic.

dimensions
The measures of an object that can be used to describe its size. For a rectangle, the dimensions are the length and width.

length / width

A. What are the dimensions of a square with an area of 1 m²?

B. Use rods as rulers to model the length and width of a square with an area of 1 m². Sketch your model.

C. What are the dimensions in centimetres of the square mosaic with an area of 1 m²? Label the dimensions for your sketch in centimetres.

D. What is the area of a square metre in square centimetres?

E. Use your answer to Part D to help you calculate the areas of the other two mosaics in square centimetres.

F. How many tiles will Rodrigo's class need to make the three mosaics? Will they have enough tiles?

Reflecting

1. If you forget how many square centimetres there are in 1 m², what picture could you sketch to help you remember?

2. Describe two methods for calculating the area of this rectangle in square centimetres.

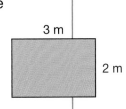

Checking

3. Rodrigo's class also plans to make a wall mosaic with an area of 0.5 m². How many 1 cm² tiles will they need for this 0.5 m² mosaic?

Practising

4. Express each area using square centimetres as the unit of measurement.
   a) 6 m²   b) 9 m²   c) 14 m²   d) 2.5 m²

5. Calculate the area of each shape in square centimetres and square metres. Show your work.

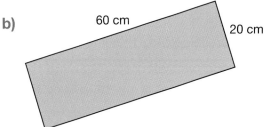

6. Alyssa made a scarf that is 125 cm by 30 cm from a 2 m² piece of fabric. How much fabric does she have left?

CHAPTER 8

Area Rule for Parallelograms

You will need
- a centimetre grid transparency
- 1 cm grid paper
- a ruler
- a protractor

Goal Develop and use a rule for calculating the area of a parallelogram.

Denise is making a stained glass pane with a pattern that includes three different parallelograms. Red glass is the most expensive, so she wants to use red for the one with the least area.

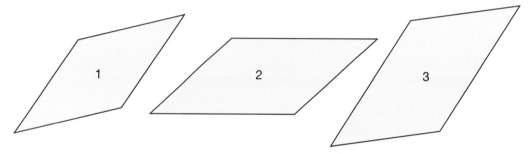

? Which parallelogram has the least area?

Denise's Solution

I don't know a rule for calculating the area of a parallelogram, but I do know the rule for calculating the area of a rectangle.

I can make a rectangle from any parallelogram.

perpendicular
At right angles

height
In a parallelogram, the height is the distance between one side of a parallelogram and the opposite side. It is measured along a line that is perpendicular to the base.

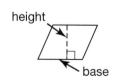

A. Trace the first parallelogram on grid paper. Draw a **perpendicular** line from the **base** to the opposite side of the parallelogram. This line is a **height**.

B. Cut out the parallelogram. Then cut the parallelogram along the height and rearrange the pieces to make a rectangle.

C. What is the area of the rectangle? What is the area of the parallelogram? How do you know?

D. Measure the base of the original parallelogram. How does it compare to the length of the rectangle?

E. Measure the height of the original parallelogram. How does it compare to the width of the rectangle?

F. How could you calculate the area of this parallelogram from the measures of the height and base?

G. Measure the base and height of the other two parallelograms. Use your answer to Part F to calculate their areas. Confirm your answers by making models on grid paper.

H. Which parallelogram should Denise make from the red glass?

Reflecting

1. a) Use words to write a rule for calculating the area of a parallelogram.
   b) How is the area rule for a parallelogram like the area rule for a rectangle? How is it different?

2. Akeem says he can't calculate the area of this parallelogram using the rule. Do you agree? Explain.

Checking

3. Denise wants to add three more parallelograms to her design.
   a) Use grid paper to draw one parallelogram for each set of measures.
   b) Estimate and then calculate the area of each parallelogram. Show your work.

Parallelogram	Base (cm)	Height (cm)
green	6	3
blue	5	4
purple	2	3

Practising

4. Calculate the area of each parallelogram. Use a ruler and protractor. Show your work.

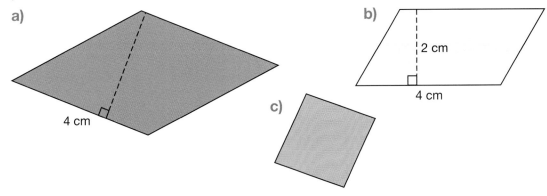

243

5. Draw two parallelograms on grid paper with an area of 18 cm². Label the length of the base and height on each.

6. This flower appliqué will be used in a quilt 12 times.
   a) What is the area of green fabric that will be needed? Show your work.
   b) What is the area of pink fabric that will be needed? Show your work.
   c) Sketch a parallelogram that has an area twice the area of one green leaf.

Mental Imagery

Estimating Area

You will need
- 1 cm square dot paper

You can estimate the area of any polygon drawn on dot paper by drawing a rectangle around the polygon.

A. How does the rectangle show that the area of the polygon must be less than 12 cm²?

B. Estimate the area of the yellow polygon.

 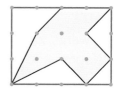

Try These

1. Copy each polygon onto dot paper. Use a rectangle to estimate the area of each polygon.

   a) 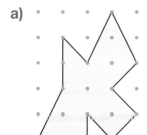 b)  c)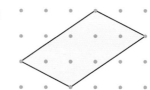

2. Draw an interesting polygon on dot paper. Show how to use a rectangle to estimate its area.

CHAPTER 8

3 Geometric Relationships

You will need
- tangram puzzle pieces
- Pairs of Congruent Triangles BLM
- scissors

Goal Identify relationships between triangles and parallelograms.

James was solving tangram puzzles when he made an interesting discovery. He could make a square by placing the two large triangles together along the longest side. He could also make a different parallelogram by joining the triangles along one of the other two sides.

? How many non-congruent parallelograms can be made with congruent triangles?

A. How many non-congruent parallelograms can be made using the two congruent isosceles right triangle tangram pieces?

B. Cut out the pairs of congruent triangles. For each type of triangle, record the number of unequal sides it has in a chart.

C. Make as many parallelograms as possible from each congruent pair. Trace your parallelograms.

D. How many non-congruent parallelograms can be made from each pair of congruent triangles? Copy and complete the table.

E. Draw your own pairs of congruent triangles and use them to repeat Part B to Part D.

Parallelograms from Triangles

Triangle	Number of unequal side lengths	Number of non-congruent parallelograms
Isosceles right triangle	2	

Reflecting

1. Can a parallelogram be made from any two congruent triangles?
2. Look at your chart. What do you notice about the number of unequal sides and the number of non-congruent parallelograms?
3. What is the relationship between the area of the parallelogram and the areas of the triangles it is made from?

CHAPTER 8

4 Area Rule for Triangles

Goal
Develop and use a rule for calculating the area of a triangle.

You will need
- 1 cm grid transparency
- 1 cm grid paper
- scissors
- a ruler
- a protractor

Raven has designed three pencil flags for her track and field team. She wants to use the one that will need the least amount of fabric to make.

? Which flag has the least area?

Raven's Solution

I don't know a rule for calculating the area of a triangle, but I do know the rule for calculating the area of a parallelogram.

I make a parallelogram using two tracings of triangle A.

I see that the height of the parallelogram is the same as the **height of the triangle** I used to make it.

The base of the parallelogram is also the same as the base of the triangle.

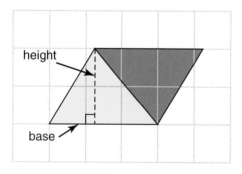

scalene acute triangle

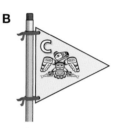
isosceles triangle

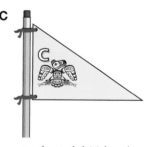
scalene right triangle

height of a triangle
A perpendicular line from a base of a triangle to the vertex opposite

A. Calculate the area of Raven's parallelogram.
B. What is the area of triangle A? Explain how you know.
C. How can you calculate the area of triangle A if you know the base and the height?

D. Measure the base and height of the other two triangles. Use your answer to Part C to calculate their areas. Make parallelograms on grid paper to confirm your answers.

E. Which triangle has the least area?

Reflecting

1. a) Write a rule for calculating the area of a triangle.
   b) How is the area rule for a triangle like the area rule for a parallelogram? How is it different?

2. Maggie says she can't calculate the area of this triangle with the rule. Do you agree? Explain.

3. How would you estimate the area of this triangle?

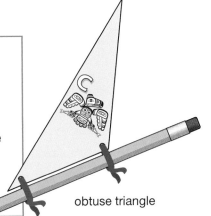
obtuse triangle

Checking

4. Raven also designed some large flags. Estimate and calculate the area of each flag. Show your work.

a)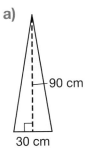
90 cm
30 cm

b)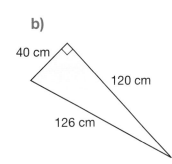
40 cm
120 cm
126 cm

c)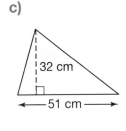
32 cm
51 cm

Practising

5. Calculate the area of each triangle. Use a ruler and a protractor. Show your work.

a)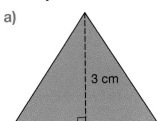
3 cm

b)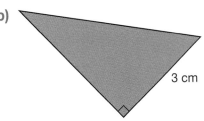
3 cm

c)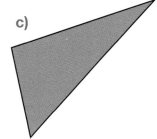

6. Draw three triangles each with an area of 6 cm². Make one a right triangle, one an acute triangle, and one an obtuse triangle. Label the length of the base and height on each triangle.

CHAPTER 8

Frequently Asked Questions

Q: Different area units may be used to measure the same area. How can you express a measurement in square metres as square centimetres?

A: Use the relationship that there are 10 000 cm² in 1 m².

> To express 11 m² in square centimetres, use a calculator to multiply by 10 000.
>
> 11 × 10 000 = 110 000
>
> You could also add 10 000 together 11 times.

Q: How are the rules for calculating the area of a rectangle, parallelogram, and triangle related?

A: The area of a rectangle is calculated by multiplying the width by the length.

Any parallelogram can be cut along a height line and be made into a rectangle with the same base and height.

Any triangle is half of a parallelogram with the same base and height.

Calculate the area.	Calculate the area.	Calculate the area.
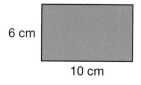	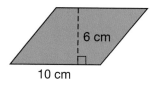	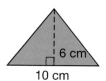
6 × 10 = 60	6 × 10 = 60	6 × 10 = 60
The area of the rectangle is 60 cm²	The area of the parallelogram is 60 cm²	half of 60 = 30
		The area of the triangle is 30 cm²

Q: How do you measure the height of a parallelogram and a triangle?

A: A parallelogram has two heights that are perpendicular lines from a base to the opposite side.

A triangle has three heights that are perpendicular lines from a base to the vertex opposite.

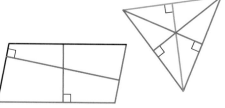

CHAPTER 8

Mid-Chapter Review

1. Express each of the following in square centimetres.
   a) 5 m^2
   b) 8 m^2
   c) 13 m^2
   d) 4.4 m^2

2. a) Carlos wants to make a skim board with an area of 1.2 m^2. He has a rectangle of plywood that has sides of 1.0 m and 1.5 m. Sketch the board.
   b) He wants to cover the surface of the board with 1 cm^2 decals. How many decals will he need?

3. Calculate the area of each parallelogram. Show your work.

   a)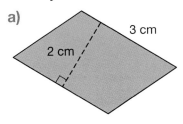
   b)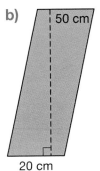
   c)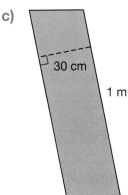

4. Draw two non-congruent parallelograms on grid paper each with an area of 16 cm^2. Label the base and height on each.

5. Calculate the area of each triangle. Use a ruler to help you. Show your work.

   a)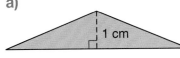
   b)
   c)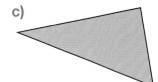

6. Draw two non-congruent triangles on grid paper each with an area of 16 cm^2. Label the base and height on each.

5 Solve Problems Using Equations

Goal Use equations to solve problems.

Kurt belongs to a theatre club. Their outdoor theatre has a rectangular seating area that is 4 m by 6 m. They want to make the seating area four times the size so they can have a larger audience.

? What dimensions could be used for the new rectangular seating area?

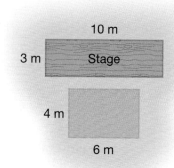

Kurt's Solution

The area of the seating rectangle is
6 m × 4 m = 24 m².
The new seating area will be four times as great, which is 24 × 4 = 96 m².
First, I'll try making the 4 m side four times as long.
4 × 4 = 16, so that side will become 16 m long.

The area 96 m² is the product of the length and width.
I can write an equation to determine the length of the other side.
■ × 16 = 96
I use guess and test and my calculator to determine the missing factor.
8 × 16 = 128   too great
5 × 16 = 80   too small
6 × 16 = 96

If one side is 16 m, then the other side needs to be 6 m. The new seating rectangle could be 6 m by 16 m but that means some people will be seated very far from the stage.

Next, I will try making the 4 m side two times longer.
2 × 4 = 8
I write an equation:
■ × 8 = 96
I know that 12 × 8 = 96
If one side is 8 m, then the other side needs to be 12 m.
The new seating rectangle could be 8 m by 12 m.

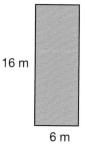

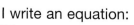
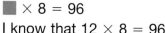

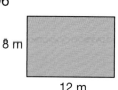

Reflecting

1. Are there other possible rectangles for the new seating area? Explain.

2. How did the strategy of using an equation help Kurt solve the problem?

Checking

3. The students made and painted a simple tree for a stage prop. They used half a can of green paint. They would like to make and paint two smaller triangle trees using most or all of the remaining paint. What base and height could they use for the new trees?

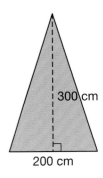

Practising

4. Mario wants to make a rectangular bulletin board for his room using 0.35 m² of cork tiles. He wants to make the width of the bulletin board 50 cm. What will the length be?

5. Nina made a sun catcher from a parallelogram with a base of 10 cm and a height of 5 cm. She wants to make another one with twice the area. What base and height could she use for the new parallelogram? List two possibilities.

6. Maria's family built a rink in their backyard last year. The rink was a 6 m by 7 m rectangle. This year they are helping to build a rink at the school. This rink will be four times the area of their backyard rink. What are three possible dimensions of a new rink? Which dimensions would you use? Explain your choice.

7. Jeff has a collection of hockey cards. His brother Steve had 10 more cards than Jeff. Steve gave Jeff all of his cards. Jeff ended up with 992 cards. How many cards did Jeff start with?

8. The office where John's father works has many cubicles in one square room. The area of the room is 144 m². A small rectangular kitchen has been built in one corner of the space. Now, the area of the office space is 120 m². Suggest two possible sets of dimensions for the new kitchen.

CHAPTER 8

6 Areas of Polygons

You will need
- pattern blocks
- a ruler
- a calculator

Goal Calculate the area of polygons by breaking them into simpler shapes.

Chandra plans to make Save Our Pond buttons of the same size and shape as a yellow hexagon pattern block. She wants to make 60 buttons.

? How many square centimetres of metal does Chandra need to make 60 buttons?

Chandra's Method

I don't know a rule for finding the area of a hexagon, but I do know the rule for finding the area of a triangle.

I can make the hexagon from green triangles.

Rebecca's Method

I don't know a rule for finding the area of a hexagon, but I do know the rule for finding the area of a parallelogram.

I can make the hexagon from blue parallelograms.

A. What is the area of a green triangle pattern block? Show your work.

B. Use the area of the green triangle to calculate the area of the yellow hexagon. Show your work.

C. What is the area of a blue parallelogram pattern block? Show your work.

D. Use the area of the blue parallelogram to calculate the area of the yellow hexagon. Show your work.

E. How many square centimetres of metal does Chandra need to make the campaign buttons?

Reflecting

1. Did you get the same area for the hexagon using both methods? Explain.

2. How did solving simpler problems help you calculate the area of a hexagon?

3. What other combinations of the blocks could you have used to find the area of the hexagon?

Checking

4. Calculate the area of each polygon in square units. Show your work.

a)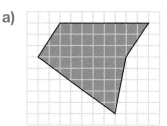

b)

Practising

5. Calculate the area of each polygon. Show your work.

a)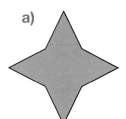

b)

6. Calculate the area of each shape. Show your work.

a)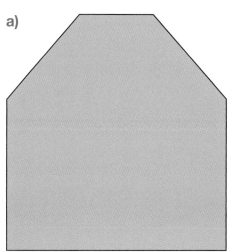

b)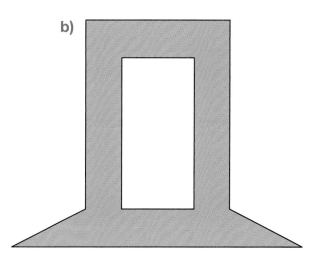

Curious Math

Changing Parallelograms

You will need
- card stock strips
- prong fasteners
- 1 cm square dot paper

1 Use the card stock strips and fasteners to make a rectangle. Trace the inside of the rectangle on centimetre dot paper. Calculate the area and the perimeter of the rectangle. Record your results in a table.

2 Keep the base of the shape fixed in position and move the top slightly to the right. What geometric shape is formed? Trace the inside of the new shape on centimetre dot paper. Calculate the area and the perimeter of the shape. Record your results.

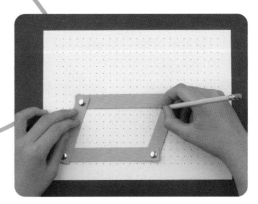

3 Repeat Step 2 three more times. Record your results in the table.

4 Look at the data in your table.
a) What remains constant?
b) What changes?
c) What shape had the greatest area?
d) As the area decreases, how does the shape change?

CHAPTER 8

Skills Bank

LESSON

1

1. Rewrite each of the following using square centimetres as the unit of measurement.
   a) 7 m²
   b) 3 m²
   c) 12 m²
   d) 16 m²
   e) 1.2 m²
   f) 7.7 m²

2

2. Copy and complete the table for these parallelograms using the measures shown.

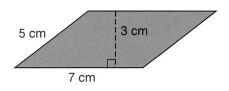
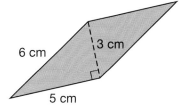
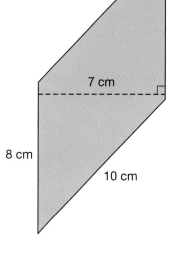

Parallelogram	Base (cm)	Height (cm)
red		
green		
blue		

3. Calculate the area of each parallelogram. Show your work.

   a)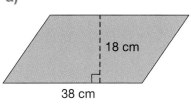
   b)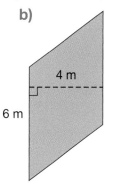
   c)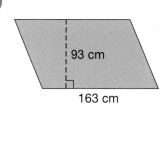

4. Draw two non-congruent parallelograms on grid paper each with an area of 24 cm².

5. Estimate and then calculate the area of each triangle. Show your work.

a)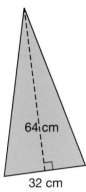

b)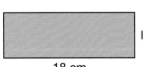

c)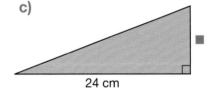

6. On centimetre grid paper, draw two triangles with each area.
   a) 20 cm²
   b) 12 cm²
   c) 25 cm²

7. Each shape has an area of 108 cm². What is the missing dimension on each shape? Show your work.

a)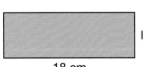

b)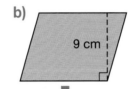

c)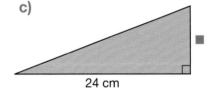

8. Anthony made a rectangular poster with a width of 32 cm and a height of 45 cm. He wants to make a new one with twice the area. What dimensions could he use? List two possibilities.

9. Calculate the area of each shape in square units.

a)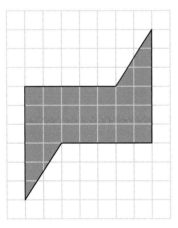

b)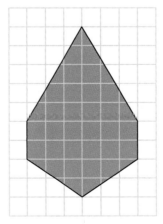

CHAPTER 8

Problem Bank

LESSON

1

1. Veronica's community group is raising money for kids in developing countries. They plan to have a marathon Scrabble® game. They have constructed a 2 m by 2 m playing board. The letter tiles are 2 cm by 2 cm. How many squares will there be on the board?

2. The town of Maplegrove has purchased a rectangular plot of land that is 0.2 km by 0.2 km to make a community garden. If the garden is divided into plots that average 50 m², how many plots will be available to rent?

2

3. Patricia designed a flag for her cabin team at camp. Redesign the flag two different ways so that the blue area becomes a parallelogram but the total area of each flag colour does not change.

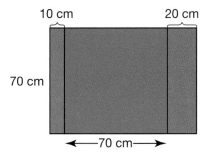

4

4. Lydia is designing an envelope for invitations she made. She plans to use two rectangular pieces with a width of 12 cm and a length of 14 cm and make the flap a triangle that goes $\frac{1}{2}$ of the way down the envelope. If she has 20 invitations, what is the area of paper she will use for the envelopes?

5. How many right triangles with different areas can you make on a 5 × 5 peg geoboard? Record your triangles on dot paper. Find one non-right triangle with the same area for each of your triangles. Record these triangles on dot paper.

6. Anna has 280 sports cards. She wants to arrange them in equal numbers on each page of a binder. List four possible ways she could arrange the cards.

7. Find the area of each tan. Make six different shapes from tans that have $\frac{1}{2}$ the area of the complete tangram. Trace each shape.

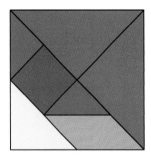

8. Patrick wants to use a tangram puzzle shape as a design on a card he is making. He plans to make this shape with gold leaf paper. What is the area of the shape? Show your work.

CHAPTER 8

Frequently Asked Questions

Q: How can you calculate a missing dimension if you know one dimension and the area of a shape?

A: Use the area rule for the shape to make an equation. Then use guess and test to calculate the missing value.

The area is 16 cm². What is the base? 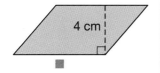	The area is 10 cm². What is the height?
■ × 4 = 16 4 × 4 = 16 The base is 4 cm.	half of 5 × ■ = 10 half of 5 × 4 = 10 The height is 4 cm.

Q: If you double the base and height of a parallelogram, what happens to the area?

A: It quadruples. The area is the product of the base and the height, so if both the base and height double, then the area will be 2 × 2 = 4 times the size.

A rectangle has a width of 3 cm and a length of 5 cm. If each dimension is doubled, what happens to the area?

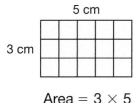

Area = 3 × 5
     = 15 cm²

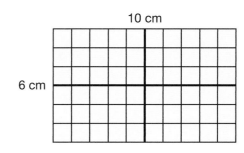

Area = 6 × 10
     = 60 cm²

The area quadruples.

Q: How can you calculate the area of a polygon?

A: Divide the polygon into triangles, rectangles, and parallelograms. Calculate the areas of the smaller shapes using area rules. Sum the areas.

Chapter Review

1. Express each of the following as square centimetres.
   a) 1 m²   b) 4 m²   c) 16 m²   d) 6.5 m²

2. Eileen made a square painting with an area of 40 000 cm². She has a 3 m by 2 m area on her bedroom wall to hang the painting on. Will the painting fit? Show your work.

3. Calculate the area of each parallelogram.

   a)
   2 cm
   5 cm

   b)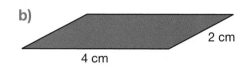
   2 cm
   4 cm

4. Without measuring or calculating, tell how you know these three parallelograms have the same area.

   2 cm   2 cm   2 cm

5. Draw two parallelograms with an area of 12 cm². Label the base and height on each one.

6. Estimate and then calculate the area of each triangle. Show your work.

   a)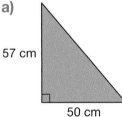
   57 cm
   50 cm

   b)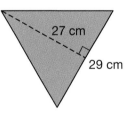
   27 cm
   29 cm

   c)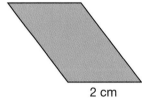
   94 cm
   78 cm
   105 cm

7. Draw two non-congruent triangles on grid paper each with an area of 10 cm². Label the base and height on each one.

8. Each shape has an area of 24 cm². Calculate the missing dimension on each one using an equation. Show your work.

a)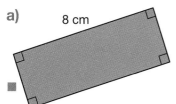
8 cm

b)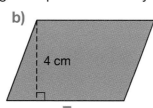
4 cm

c)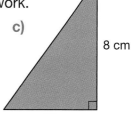
8 cm

9. Mika's family is making concrete stones with footprints on them for a walkway. They started by making the stones with the children's footprints. They plan to make the stones for the adult footprints with twice the area. What base and height could they use for the adult stones? List two possibilities.

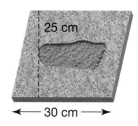
25 cm
30 cm

10. Calculate the area of each shape in square units. Show your work.

a)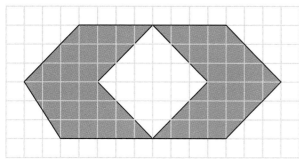

b)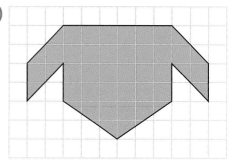

11. Calculate the area of each shape. Show your work.

a)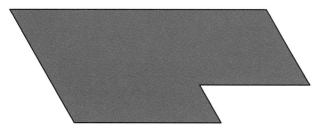

b)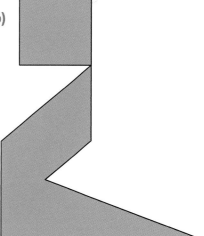

CHAPTER 8

Chapter Task

Design a Placemat and Napkin Set

Angele is making a set of placemats and napkins as a gift. She has designed a pattern for them using geometric shapes. She will make the placemats using different colours of fabric.

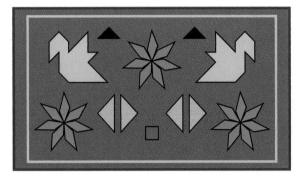

? **How can you design a set of placemats and napkins and calculate the area of each colour?**

A. Most placemats have an area of between 1600 cm^2 and 2000 cm^2. The width of a placemat is at least 35 cm. Decide on dimensions for your placemat.

B. Make a model of your placemat from paper. Use tangram pieces to create a design for your placemat. Use each shape at least once. Include a shape in your design that is made from more than one tangram piece.

C. Colour your design, including the background, in at least four different colours.

D. Estimate the area covered by each colour.

E. Calculate the area covered by each colour.

F. If you make your napkins rectangles that have half the area of your placemats, what dimensions could you use?

G. Decide on the number of placemats and napkins in your set. For each colour, what is the total amount of fabric you will need?

Task Checklist
- ✓ Did you measure the dimensions accurately?
- ✓ Did you use the correct area rules?
- ✓ Did you show all of your work for your calculations?
- ✓ Did you include appropriate units?

CHAPTER 9

Multiplying Decimals

Using scale

Goals

You will be able to

- **estimate products involving decimals and explain your strategies**
- **multiply decimals by 10, 100, and 1000 using mental math**
- **multiply whole numbers by numbers with one decimal place**
- **multiply whole numbers by 0.1, 0.01, and 0.001**
- **select an appropriate multiplication method**

CHAPTER 9

Getting Started

Downloading Songs

Emilio found two Web sites for downloading songs and albums. He spent almost all of his savings of $40 on downloads. His mother agreed to pay the tax.

? **How many songs or albums might Emilio have downloaded?**

A. Suppose Emilio downloaded only songs, but no albums. How do you know that he downloaded about 40 songs?

B. About how many more songs could Emilio have purchased at the $0.88 site than at the $0.99 site with his $40? How do you know?

C. Suppose Emilio downloaded only albums. What could his change from $40 be if he bought three albums? List two possibilities.

D. What is the least number of purchases Emilio could have made and have less than $1 left?

E. List three possible song/album combinations that Emilio could have purchased and have less than $1 left.

F. Create and solve your own problem about the cost of downloading music.

Do You Remember?

1. What decimal does the shaded area of each model represent?

   a)   b)   c)

2. What multiplication does the model represent?

   a) 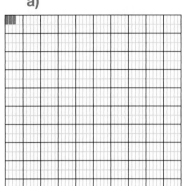   b)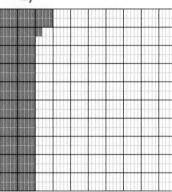

3. Estimate each product.
   a) 6 × 5.2   b) 4 × 3.9   c) 8 × 5.8

4. Calculate.
   a) 6 × 303   c) 7 × 336   e) 10 × 25.3
   b) 5 × 412   d) 9 × 534   f) 100 × 0.56

5. Choose two calculations to perform using mental math. Calculate those two.
   A. 6 × 427   B. 6 × 15   C. 10 × 345   D. 2 × 43

CHAPTER 9

1 Estimating Products

 Goal Estimate products of decimal tenths and money amounts using a variety of strategies.

James and his family are buying groceries.

? About how much will the groceries cost?

 James's Method

I'll estimate the cost of the salmon by multiplying 2 × $18. I know that estimate will be high.

A. What is James's estimate? How did he know it would be high?

B. Use each calculation to estimate the cost of the salmon steaks. Explain why each is reasonable.
- 1 × $20
- 1 × $18 + $\frac{1}{3}$ of $18

C. Which of the three estimates in Parts A and B do you think is closest to the actual cost of the salmon steaks? Explain.

D. Estimate the cost of each of the other grocery items in two ways. For each item, tell which estimate you think is closer to the actual price and why.

E. Give an example of three items with different masses and different prices that would cost about the same.

Reflecting

1. When you estimate the price of 0.3 kg of something, why might you want to estimate the price of 1 kg as a multiple of 3?
2. a) Estimate the cost of each amount of nuts.
   b) Why might you have estimated each cost in a different way?
3. Why is there always more than one way to estimate products involving decimals? Explain, using $3.8 \times \$11.49$ as an example.

Mental Math

Multiplying by 5 and 50

You can multiply a number by 5 by thinking of 10 instead.

You can multiply a number by 50 by thinking of 100 instead.

Qi's Method

I can multiply 28 by 5 by renaming 28×5 as $28 \times 10 \div 2$.

$28 \times 10 = 280$
$280 \div 2 = 140$
So, $28 \times 5 = 140$

Rodrigo's Method

I can multiply 28 by 50 by renaming 28×50 as $28 \times 100 \div 2$.

$28 \times 100 = 2800$
$2800 \div 2 = 1400$
So, $28 \times 50 = 1400$

A. Use the answer to 28×5 to calculate 29×5. Show your work.

Try These

1. Complete each equation.
   a) $5 \times 16 = \blacksquare$
   b) $32 \times 5 = \blacksquare$
   c) $5 \times 26 = \blacksquare$
   d) $48 \times 5 = \blacksquare$
   e) $50 \times 26 = \blacksquare$
   f) $64 \times 50 = \blacksquare$
   g) $50 \times 22 = \blacksquare$
   h) $50 \times 84 = \blacksquare$

CHAPTER 9

2 Multiplying by 1000 and 10 000

You will need
- counters
- a place value chart

Goal: Multiply decimal tenths, hundredths, and thousandths by 1000 and 10 000.

Isabella did a career project about architects. She built a scale model of First Canadian Place, one of the tallest office buildings in Toronto.

The real building is 1000 times as high as the model. Isabella lives 10 times as far from school as the building is high.

? How far does Isabella live from school?

Isabella's Strategy

29 cm is the same as 0.29 m, so 29.8 cm is 0.298 m.

I need to multiply 0.298 by 1000 to determine how many metres high First Canadian Place is.

I'll model this on the place value chart.

Thousands	Hundreds	Tens	Ones (metre)	Tenths	Hundredths (cm)	Thousandths (mm)
				●●	●●●●●●●●●	●●●●●●●●

I'll have to move the counters to the left.

A. Why did Isabella say that 29.8 cm is the same as 0.298 m?

B. How many columns to the left would you move each counter to multiply its value by 1000? Why?

C. How many metres tall is First Canadian Place?

D. How far from school does Isabella live?

Reflecting

1. Why did Isabella multiply 0.298 by 1000 to determine the height of the building in metres?
2. Why was calculating her distance from school like multiplying 0.298 by 10 000?
3. Describe a rule for multiplying a decimal by 1000 or 10 000. How is it like multiplying by 10 or 100?

Checking

4. The CN Tower is 1000 times as high as a model that is 55.3 cm high. How many metres tall is the CN Tower?

Practising

5. Calculate.
   a) 1000 × 2.344
   b) 1000 × 13.24
   c) 1000 × 3.2
   d) 10 000 × 12.456
   e) 10 000 × 0.45
   f) 10 000 × 12.48

6. How many metres apart are the items in each pair?

   a)  31.213 km

   b)  2.14 km

   c)  4.7 km

7. One pack of paper is 1.4 cm thick. How many metres high would a stack of 10 000 packs be? How do you know?

8. Multiply 3.58 by 1000. How could you have predicted that the answer would have a zero in the ones place?

9. Gas cost 79.9¢ a litre. Daniel's dad used 1000 L of gas every 4 months. About how much did he spend on gas in 1 year?

10. Amanda walks 3.2 km a day. About how far will she walk in 3 years?

CHAPTER 9

3 Multiplying Tenths by Whole Numbers

You will need
- base ten blocks
- a decimal place value chart

Goal: Multiply decimal tenths by whole numbers using models, drawings, and symbols.

Akeem can run 100 m in 12.4 s.

? How long would it take Akeem to run 300 m at that speed?

Akeem's Method

I'll multiply to figure out the distance.

I'll use base ten blocks. The rod | will represent 1 s.

I'll model 3 groups of 12.4.

Tens	Ones •	Tenths
		• • • • • • • • • • • •

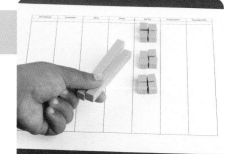

```
  1
 12.4        4 tenths
× 3          × 3
────        ────────
  .2         12 tenths
```

A. How much does a small cube represent in Akeem's model? How much does a flat represent?

B. How do you know that 3 × 12.4 is a little more than 36?

C. Copy and complete Akeem's work. Show all of the blocks and record the multiplication with symbols. How long would it take him to run 300 m at that speed?

Reflecting

1. Was using the rod to represent a one a good decision? Explain.

2. Why does the product include tenths when you multiply a decimal with tenths by a whole number?

3. How is multiplying 3 × 12.4 like multiplying 3 × 124? How is it different?

Checking

4. Calculate using base ten blocks as a model. Sketch your model. 5 × 26.7 = ■

5. Jane walks 2.6 km each day. How far does she walk in 1 week?

Practising

6. What is the perimeter of this patio, which is in the shape of a regular hexagon?

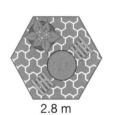

2.8 m

7. Elizabeth can run 200 m in 25.7 s. How long would it take her to run 1000 m at that speed?

8. David and his friends are making lasagna for a school dinner. There will be 400 people at the dinner. The recipe for 100 people uses 5.6 kg of hamburger, 3.4 L of cottage cheese, and 2.7 kg of pasta.
   a) How much of each ingredient is needed?
   b) One kilogram of hamburger costs $9. What is the total cost of the hamburger that is needed?

9. Make up a problem in which you multiply 5 × 23.6. Solve it.

10. Stephanie entered the digits 3, 4, 5, and 6 in the spaces, not necessarily in that order, and multiplied. ■ × ■■.■ Zachary entered them in a different order and multiplied. Stephanie's product was 12.8 greater than Zachary's product. How did each student arrange the digits?

11. To multiply 7 × 3.7, Jane writes 7 × 37 tenths = 259 tenths.

    Then she writes 259 tenths = 25.9.
    Is Jane correct? Explain.

CHAPTER 9

Frequently Asked Questions

Q: How do you estimate the product of two decimals?

A: Replace each decimal with a convenient whole number and multiply them. You can try to get closer to the actual answer by adding or subtracting an appropriate amount. For example, 4.3×6.12 is about $4 \times 6 = 24$.

To get closer, think: 4.3×6.12 is about $4 \times 6 +$ another third of 6 since 0.3 is about $\frac{1}{3}$.

That estimate is $24 + 2 = 26$.

4.8×6.9 is about $5 \times 7 = 35$. Subtract a bit since both numbers were increased. An estimate might be 33.

Q: How do you multiply a decimal by 10 000?

A: Move each digit four spaces to the left on the place value chart. The thousandths digit becomes the tens digit and the ones digit becomes the ten thousands digit. For example:
$10\,000 \times 3.056 = 30\,560$

Ten Thousands	Thousands	Hundreds	Tens	Ones •	Tenths	Hundredths	Thousandths
				3		5	6

Ten Thousands	Thousands	Hundreds	Tens	Ones •	Tenths	Hundredths	Thousandths
3		5	6				

Q: How do you multiply a decimal by a single-digit whole number?

A: Multiply each part of the decimal by the whole number. You can also write the decimal as a whole number of tenths and then rewrite the product as a decimal. For example, for 3×23.4:

```
  1
 23.4      234 tenths
× 3        × 3
─────      ──────────────
 70.2      702 tenths = 70.2
```

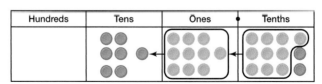

CHAPTER 9

Mid-Chapter Review

1. Estimate the cost for each in two different ways.

   a)
   b)

2. Estimate how much more the salmon costs than the hamburger.

   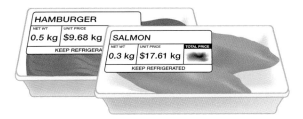

3. Dave noticed that two exits on the highway were 2.4 km apart. How many metres apart were they?

4. When you multiply a decimal, like 3.125, by 1000, the answer is a whole number. But when you multiply the decimal by 100, the answer is not a whole number. Why would you expect this?

5. Calculate.
   a) 1000 × 5.124
   b) 1000 × 9.03
   c) 10 000 × 37.8
   d) 10 000 × 0.042

6. Use base ten blocks to model each product. Sketch your model and copy and complete each equation.
   a) 6 × 7.1 = ■   b) 4 × 3.4 = ■   c) 5 × 37.7 = ■

7. A case of apple juice holds eight cans. Each can contains 1.4 L of juice.
   a) How do you know that the total amount of juice is less than 12 L?
   b) How many litres of juice are in the case?

CHAPTER 9

4 Multiplying by 0.1, 0.01, or 0.001

You will need
- a decimal place value chart
- counters

Goal Multiply by 0.1, 0.01, or 0.001 using mental math.

Raven measured the distance between two desks with base ten rods. They are 22 rods apart.

? How many metres apart are the desks?

Raven's Method

Each rod is 10 cm long. I can write 10 cm as 0.1 m.

22 rods is 22 tenths of a metre.
22 tenths can be written as 22 × 0.1.

22 tenths = 2 ones and 2 tenths.
22 × 0.1 = 2.2

The digits for 22 × 0.1 are still 2 and 2, but they are in different place values in the place value chart. The desks are 2.2 m apart.

Tens	Ones	Tenths	Hundredths	Thousandths
		●● ●●●●● ●●●●● ●●●●● ●●●●●		

Tens	Ones	Tenths	Hundredths	Thousandths
	●●	●●		

Reflecting

1. Why did Raven write 22 tenths as 22 × 0.1?
2. a) Why can you write 22 × 0.01 as 0.22?
   b) Why can you write 22 × 0.001 as 0.022?
3. Create a rule for multiplying by 0.1, by 0.01, and by 0.001.

274 NEL

Checking

4. What is each length?
   a) 15 rods, each 0.1 m long, in metres
   b) a line of 27 cubes, each 0.01 m long, in metres
   c) a line of 345 sticks, each 0.001 km long, in kilometres

Practising

5. In Kuwait, people use dinars instead of dollars. A dinar is worth 0.001 fils. How many dinars is 47 fils worth?

6. a) Use a decimal to describe the part of a kilogram that 1 g represents.
   b) A bag of popcorn is 99 g. How many kilograms is this?

7. a) Complete the pattern:
   316 × 100 = ■
   316 × 10 = ■
   316 × 1 = ■
   316 × 0.1 = ■
   316 × 0.01 = ■
   316 × 0.001 = ■
   b) Describe the pattern.

8. Which products are greater than 1? How do you know?
   A. 75 × 0.1   B. 318 × 0.001   C. 214 × 0.01

9. In which place value in the chart will the digit 3 appear after each calculation?

Hundreds	Tens	Ones	•	Tenths	Hundredths	Thousandths

   a) 342 × 0.01   b) 35 × 0.001   c) 4325 × 0.1

10. Write a whole number less than 100 with 7 as one of the digits. Multiply that number by 0.1, by 0.01, or by 0.001, so that the 7 ends up in the hundredths place. Record the multiplication.

11. Amber said that to multiply 3000 × 0.001 you can divide 3000 by 1000. Is this also true for other numbers? Explain.

CHAPTER 9

5 Multiplying Multiples of Ten by Tenths

You will need
- a place value chart

Goal: Multiply to calculate the decimal portion of a multiple of 10.

The Grade 6 students from Maple School gathered these data about youth fitness in Watertown.

- **Exercise:** The number of students who exercise regularly is 10 times the number of students in Maple School.
- **Skating:** The number of students who skate is the same as the number of students in Maple School.
- **Volleyball:** One tenth (0.1) of the students in Maple School play volleyball.
- **Basketball:** Three tenths (0.3) of the students in Maple School play basketball.

? How many students do each activity?

Jorge's Method

Exercise: 10 times as many as 650 is like having 10 groups of 650. I can multiply 10 × 650 to calculate the number of students who exercise regularly. 10 × 650 = 6500

Skating: One group of 650 is 1 × 650. I can multiply 1 × 650 to determine the number of students who skate. The digits from 10 × 650 move to the right one place for 1 × 650.

Volleyball: One tenth of 650 is 0.1 groups of 650. I can multiply 0.1 × 650 to determine the number of students who play volleyball. The digits of 1 × 650 move to the right one place.
0.1 × 650 = 65

Basketball: Since 0.3 is 3 times as much as 0.1, I multiply 3 × 65 to determine the number of students who play basketball.
0.3 × 650 = 3 × 65
= 195
195 students play basketball.

		Thousands	Hundreds	Tens	Ones
Exercise	10 × 650 =	6	5	0	0
Skating	1 × 650 =		6	5	0
Volleyball	0.1 × 650 =			6	5

Reflecting

1. Does the pattern in the place value chart make sense? Explain.

2. Why is multiplying a number by 0.1 the same as taking one tenth of it or dividing it by 10?

3. Could you calculate 0.6 × 700 by calculating 0.1 × 700 and then multiplying by 6? Explain.

Checking

4. 0.1 of the 300 students in Leigh's school are going on a trip to Ottawa.
    a) Multiply to calculate how many students are going on the trip. Show your work.
    b) Divide to calculate how many students are going on the trip. Show your work.

Practising

5. Calculate.
    a) 0.1 × 480   b) 0.6 × 220   c) 0.2 × 1230   d) 0.9 × 140

6. 0.6 of the 450 students in a school voted in the student elections.
    a) How many students voted?
    b) How do you know that the answer is reasonable?

7. At a birthday party, 10 children ate 0.7 of a 2000 mL tub of ice cream.
    a) What volume of ice cream did the children eat altogether?
    b) Each child ate the same amount of ice cream. How much did each child eat?

8. 0.5 of the 760 students in a school are participating in a fitness project. How many students are participating?
    Begin the solution of this problem in each one of the following ways.
    a) Represent 760 with counters on a place value chart.
    b) Multiply 760 by 0.1 using mental math.
    c) Multiply 760 by 5 on paper.
    d) Write 0.5 as $\frac{1}{2}$.

277

CHAPTER 9

6 Communicate About Problem Solving

You will need
- a calculator

Goal: Explain how to solve problems involving decimal multiplication.

Chandra explained how she solved this problem. "Leanne has saved $120. She promised her mother not to spend more than $\frac{6}{10}$ of her savings on a new bike. What is the least she could have left?"

Denise gave Chandra some advice.

Chandra's Explanation

$\frac{6}{10}$ is 0.6.

$0.1 \times 120 = 12$ ← Why did you multiply 0.1×120?

$6 \times \$12 = \$72$ ← Why did you multiply 6×12?

$\$120 - \$72 = \$48$

Did you show the steps of the problem solving process?

Could you draw a picture to make this clearer?

? How can you improve Chandra's explanation?

A. Improve Chandra's explanation by responding to Denise's questions.

B. How else could Chandra improve her explanation?

Communication Checklist
- ☑ Did you model the problem-solving process?
- ☑ Did you show all your steps?
- ☑ Did you use a model or diagram?
- ☑ Did you explain your thinking?
- ☑ Did you clearly state your solution?

Reflecting

1. Do you think Chandra understood how to solve the problem? Why?
2. Did Chandra communicate well? Explain your opinion.

Checking

3. Kurt has earned $50 by walking dogs. He will save 0.4 of that and spend the rest. He decided to determine how much he could spend. He explained his thinking.

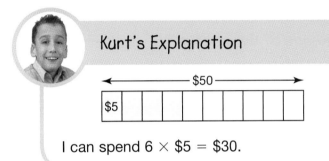

Kurt's Explanation

I can spend 6 × $5 = $30.

a) Identify at least one strength in Kurt's explanation. Use the Communication Checklist.
b) Identify at least one weakness of his explanation.
c) Improve his explanation.

Practising

4. The mass of a penny is about 0.002 kg. What is the mass of 900 pennies? Explain your thinking.

5. Jeff's mass is 48 kg. The mass of his backpack should not be more than $\frac{1}{6}$ of Jeff's mass.

   Jeff has four textbooks in his backpack, each with a mass of 0.9 kg. How much more mass can Jeff carry safely?

6. The world's heaviest pumpkin had a mass of 606.7 kg. A typical pumpkin has a mass of 9.1 kg. About how many typical pumpkins would balance the heaviest one? Explain your thinking.

7 Choosing a Multiplication Method

You will need
- a calculator

Goal Justify the choice of a multiplication method.

Li Ming has made a poster for her school concert and is going to make copies.

Photocopier	1	2	3	4	5
Time to print one copy	0.8 s	0.6 s	2.5 s	2.9 s	0.45 s

? How long would it take to print copies of the poster?

A. How long would it take each photocopier to print 100 copies?

B. How could you use mental math to answer Part A?

C. How long would it take each photocopier to print nine copies?

D. For which of the calculations in Part C did you use pencil and paper? Explain your choice.

E. For which of the calculations in Part C did you use a calculator? Explain your choice.

F. "Use mental math to calculate the length of time Photocopier 1 takes to print ■ copies." Give four possible numbers that you could put in the blank. Explain your choices.

G. Make up three problems about printing copies on one or more of the photocopiers in the chart. Use mental math, pencil and paper, or a calculator to solve each problem. In each case, show your work and explain why you chose the method you did.

Reflecting

1. How do you decide when to use mental math to calculate a product involving decimals?

2. How do you decide whether to solve a multiplication problem using pencil and paper or a calculator if one of the numbers involves decimals?

Curious Math

Decimal Equivalents

If you know the decimal equivalent for one fraction, it can help you to calculate the decimal equivalents for other fractions.

For example, divide 1 by 8 on a calculator to calculate the decimal equivalent for $\frac{1}{8}$.

$1 \div 8 = 0.125$

To calculate the decimal equivalent for $\frac{2}{8}$, multiply 2×0.125, which is 0.250.

This makes sense, since $0.250 = 0.25$, which is $\frac{1}{4}$ and $\frac{2}{8} = \frac{1}{4}$.

1. Calculate the decimal equivalents for $\frac{3}{8}, \frac{4}{8}, \frac{5}{8}, \frac{6}{8}, \frac{7}{8}$, and $\frac{8}{8}$ by multiplying each numerator by 0.125 on your calculator. How do the values for $\frac{4}{8}$ and $\frac{6}{8}$ help to confirm that your answers make sense?

2. Calculate the decimal equivalent for $\frac{1}{9}$. Use that value to calculate the decimal equivalents for $\frac{2}{9}, \frac{3}{9}, \frac{4}{9}, \frac{5}{9}, \frac{6}{9}, \frac{7}{9}, \frac{8}{9}$, and $\frac{9}{9}$. What do you notice?

Math Game

Race to 50

You will need
- a deck of 40 number cards (4 each of the digits from 0–9)

Number of players: two or more

How to play: Multiply a decimal number by a whole number and determine the tenths digit of the product.

Step 1   Shuffle the cards.
         Deal four cards to each player.

Step 2   Each player arranges three cards into a number between 10 and 100.

Step 3   Multiply this number by the digit on the fourth card. The digit in the tenths place of the product is your score.

Step 4   Add your score from each round.

The first person to reach 50 points wins.

Tara's Turn

I could multiply 2 × 25.7.

$\boxed{2} \times \boxed{2}\boxed{5} \cdot \boxed{7} = 51.4$

The answer is 51.4, so I'd get 4 points.

I could multiply 5 × 27.2.

$\boxed{5} \times \boxed{2}\boxed{7} \cdot \boxed{2} = 136.0$

The answer is 136.0, so I'd get 0 points.

I'll choose the first way, because that gives me more points.

CHAPTER 9

Skills Bank

LESSON 1

1. Estimate. Which products are between 40 and 50?
   - A. 6.87 × 6
   - B. 7.7 × 4
   - C. 13.3 × 4
   - D. 4.9 × 9
   - E. 8.1 × 3
   - F. 5.1 × 9
   - G. 18.6 × 2
   - H. 75.9 × 8
   - I. 10.01 × 4

2. Estimate each product as a whole number.
   - a) 4.12 × 3.9
   - b) 11.3 × 4.1
   - c) 2.9 × 3.83
   - d) 7.7 × $6.79
   - e) 6.4 × $5.49
   - f) 8.8 × $3.29
   - g) 1.6 × 19.4
   - h) $5.08 × 61
   - i) 45.2 × $7.21

3. Tell whether each estimate is probably high or low. Explain your answer.
   - a) 7.41 × 3
     Estimate: 21
   - b) 8.9 × 3.9
     Estimate: 36
   - c) 6.5 × 5.1
     Estimate: 35
   - d) 7.4 × 3.39
     Estimate: 22
   - e) 10.1 × 8.3
     Estimate: 83
   - f) 19.2 × 4
     Estimate: 80

LESSON 2

4. A model of the Taipei tower is 50.9 cm high. The Taipei tower is 1000 times as high as the model. How many metres high is the Taipei tower?

5. Calculate each product.
   - a) 1000 × 0.004 = ■
   - b) 1000 × 17.242 = ■
   - c) 1000 × 3.4 = ■
   - d) 10 000 × 3.18 = ■
   - e) 10 000 × 0.06 = ■
   - f) 10 000 × 1.908 = ■

6. Which product is greater? How much greater?
   - a) 1000 × 0.45 or 10 000 × 0.018
   - b) 1000 × 1.37 or 1000 × 0.9
   - c) 10 000 × 0.07 or 10 000 × 0.044

7. Do the multiplication that each model represents. Show your work.

a)
Tens	Ones	Tenths
6 squares	2 rods	6 dots
(3 rows of 2)	(3 rows of 2)	(3 rows of 2)

b)
Tens	Ones	Tenths
5 squares		10 dots

c)
Tens	Ones	Tenths
8 squares	12 rods	36 dots

8. Calculate. Use base ten blocks as a model. Sketch your models and copy and complete each equation.
   a) 8 × 0.7 = ■
   b) 5 × 2.4 = ■
   c) 3 × 0.8 = ■
   d) 4 × 8.1 = ■
   e) 7 × 12.3 = ■
   f) 8 × 5.9 = ■
   g) 2 × 8.2 = ■
   h) 5 × 1.3 = ■
   i) 4 × 3.8 = ■

9. a) Calculate the first three products in the pattern.
   5 × 1.3 = ■
   5 × 1.4 = ■
   5 × 1.5 = ■
   b) Write the next three products.
   c) Predict what the 20th product in the pattern will be.

10. Write each length in metres.
    a) 1728 cm = ■ m
    b) 1728 mm = ■ m

11. In which place value is the number 4 in each product? Explain.
    a) 324 × 0.1
    b) 1542 × 0.01
    c) 14 231 × 0.001
    d) 54 × 0.01

12. Calculate.
    a) 782 × 0.1 = ■
    b) 403 × 0.01 = ■
    c) 345 × 0.001 = ■
    d) 1157 × 0.01 = ■
    e) 38 × 0.1 = ■
    f) 5430 × 0.01 = ■
    g) 85 × 0.001 = ■
    h) 406 × 0.1 = ■
    i) 310 × 0.001 = ■
    j) 10 000 × 0.001 = ■

13. Calculate.
    a) 0.4 × 50 = ■
    b) 0.8 × 600 = ■
    c) 0.7 × 200 = ■
    d) 0.5 × 450 = ■
    e) 0.9 × 250 = ■
    f) 0.2 × 320 = ■

14. Calculate.
    a) 0.2 × 360 = ■
    b) 0.3 × 40 = ■
    c) 0.7 × 220 = ■
    d) 0.9 × 600 = ■
    e) 0.6 × 160 = ■
    f) 0.5 × 1220 = ■

15. 0.3 of the 800 students in a school participate in extracurricular activities. How many students participate?

16. 0.7 of the 640 students in a school have pets. How many students have pets?

17. Which calculations can you do mentally? Record the product for each of those.
    A. 832 × 0.001
    B. 1000 × 0.45
    C. 76 × 0.35
    D. 11 × 4.3
    E. 5 × 6.24
    F. 20 × 2.5
    G. 0.83 × 40
    H. 6.2 × 3.74
    I. 0.3 × 21

18. Which calculations from Question 17 would you do with pencil and paper? Record the product for each of those.

CHAPTER 9

Problem Bank

LESSON

1
1. If 2.2 kg of potatoes costs $3.99, then about how much does 5 kg of potatoes cost?

2
2. A photocopying service charges 2.7¢ for each copy it prints. How much would it charge to print 1000 photocopies?

3. A lumber company produced 1000 wooden boards. Unfortunately, each board was 2.1 mm too long and that amount had to be cut off from each board! In all, how many metres of wood were cut off?

3
4. When Justin was born, he was 45.7 cm long. Now he is four times as tall.
   a) How tall is he now?
   b) How old do you think he might be? Explain.

4
5. 0.4 of the people in Greenton and 0.3 of the people in Woodton are participating in a recycling competition. Each town's population is estimated to the nearest hundred. In total, 1230 people are participating. What might the population of each town be? List two possibilities.

6. 0.6 of the adults in Smalltown are married. 1200 adults are married. How many adults live in Smalltown?

5
7. Place the digits so that the product is as close to 500 as possible.

   0.■ × ■■■

CHAPTER 9

Frequently Asked Questions

Q: How can you use mental math to multiply a number by 0.1, by 0.01, or by 0.001?

A: Use place values. The digits in the result are the same as the digits you start with. They just move to the right one, two, or three places. For example, when you multiply 107 by 0.1, 0.01, or 0.001, you have 107 tenths, hundredths, or thousandths.
$107 \times 0.1 = 107$ tenths, or 10.7
$107 \times 0.01 = 107$ hundredths, or 1.07
$107 \times 0.001 = 107$ thousandths, or 0.107

Q: What does an expression like 0.3×200 mean and how do you calculate it?

A: 0.3×200 means $\frac{3}{10}$ of 200.
This is 3 times as much as $\frac{1}{10}$ of 200.
$\frac{1}{10}$ of 200 is $0.1 \times 200 = 20$.
So $\frac{3}{10}$ of 200 is $3 \times 20 = 60$.
$0.3 \times 200 = 60$. This makes sense, since 0.3×200 is more than $\frac{1}{4}$ of 200, but less than $\frac{1}{2}$ of 200.

Q: What decimal products can you usually calculate using mental math?

A: You can multiply decimals by 10, 100, 1000, or 10 000 using mental math. For example, $0.256 \times 1000 = 256$.

You can multiply numbers by 0.1, 0.01, 0.001 using mental math by moving digits appropriately. For example, $0.01 \times 356 = 3.56$.

Some other products based on multiplication facts are also easy to calculate using mental math. For example, 20×0.007 is just twice as much as 10×0.007, so it's 2×0.07. 0.07 is 7 hundredths, so 2×7 hundreds is 14 hundredths. That's 0.14.

CHAPTER 9

Chapter Review

LESSON

1

1. Estimate the price of the food.

   a)

   b)

2. Estimate each product using whole numbers. Explain your strategy for one of your estimates.

   a) 3.1 × 14.2   b) 4.6 × 8.9

2

3. It is 2.8 km from Meg's house to her cousin's house. How many metres apart are their houses?

4. Calculate each product.
   a) 1000 × 1.83 = ■
   b) 10 000 × 0.026 = ■
   c) 10 000 × 2.1 = ■

5. 10 000 × 3.456 = 1000 × ■
   What number goes in the box?

3

6. Calculate each product using base ten blocks. Sketch your model.
   a) 4 × 1.8 = ■   b) 5 × 4.3 = ■   c) 9 × 2.4 = ■

7. Dave drinks 0.8 L of milk a day.
   a) How many litres of milk does he drink in a week?
   b) How many 2 L containers of milk should he buy each week?

8. Courtney walks 2.6 km door to door to get to school. What distance does she cover in 4 days if she walks both ways each day?

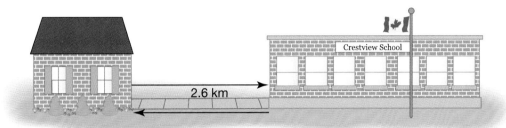

9. Calculate 3482 × 0.001. Explain what you did.

10. In which place value of the product is the digit 8?
    182 × 0.01

11. Why is the product 0.001 multiplied by a number always less than that number?

12. How much greater is 0.6 × 500 than 0.5 × 600? How do you know?

13. How are these products related?
    0.4 × 200   0.4 × 100   0.2 × 400   0.1 × 200

14. Draw a picture to show why 0.1 × 350 = 35.

15. Solve the problem and explain your thinking:
    A recipe calls for 0.3 L of juice for one batch.
    You have 5 L of juice.
    How much juice is left over after you've made seven batches of the recipe?

16. Explain why a store earns $4760 if it sells 1000 copies of a book that costs $4.76.

17. Which digits can go in the blanks so that you can calculate ■ × 1■.■ using mental math. Do the calculation, explaining what you did.

18. Explain how you would solve 423 × 0.9 using mental math.

CHAPTER 9

Chapter Task

Growing Up

A newborn baby has a mass of 3.6 kg. The baby is expected to gain an average of 0.2 kg each week for the first 6 months.

A certain breed of puppy is 0.4 kg at birth, but is expected to gain about 0.6 kg each week in the first 4 weeks. After that, it gains 1.2 kg each week for the next 4 weeks. Then it gains 1.5 kg each week for the next 4 months.

? How do the masses of the baby and the puppy compare at different ages?

A. Calculate the baby's expected mass at these ages:
 • 4 weeks • 8 weeks • 12 weeks • 24 weeks
 How do you know that your results are reasonable?

B. Calculate the puppy's expected mass at the same ages as in Part A.

C. Which of the calculations in Parts A and B did you perform using mental math? Explain your choice.

D. At about what age is the birth mass doubled for the baby? for the puppy? Explain.

E. Estimate the baby's mass at 12 weeks as a decimal 0.■ of its mass at 24 weeks.

F. Estimate the puppy's mass at 12 weeks as a decimal 0.■ of its mass at 24 weeks.

G. Describe the difference in how babies and puppies grow.

Task Checklist
- ✓ Did you use efficient calculation procedures?
- ✓ Did you explain your thinking?
- ✓ Did you support your conclusions?
- ✓ Did you estimate when it was appropriate?

CHAPTER 10

Dividing Decimals

Goals

You will be able to

- estimate the quotient when dividing a decimal
- divide whole numbers and decimals by whole numbers
- use division and other operations to solve multi-step problems
- solve problems by working backward

Between the hurdles

CHAPTER 10

Getting Started

You will need
- a measuring tape or a metre stick
- string

Ancient Length Measurements

For a social studies project, Rebecca investigated some units of length that people have used. She made a chart to show the approximate length of each unit in centimetres.

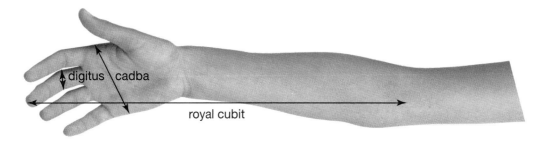

Unit	digitus	sun	cadba	foot	charal	royal cubit	pole
Approximate length in centimetres	2	3	8	31	26	52	65

? **How can you change your measurements into an ancient unit of length?**

A. Measure your height to the nearest centimetre. Choose a unit from the chart that is less than 10 cm. Show how to change your height into that ancient unit of length.

B. Measure the length of the classroom to the nearest centimetre. Choose one unit from the chart that is greater than 10 cm. Show how to change the classroom length into that ancient unit of length.

C. Explain why the measurement of the classroom length in royal cubits will be half of that for the classroom length in charals.

D. Explain how you know that your answers to Parts A and B are reasonable.

E. Explain how you can use division to change a length measurement in centimetres into an ancient unit of length.

F. Use string to create your own unit of length. Measure the length of your unit in centimetres. Make up a name for your unit. Show how to change a length that you have measured in centimetres into your unit of length.

Do You Remember?

1. Calculate. Show your work.
   a) 159 ÷ 4 = ■
   b) 2907 ÷ 6 = ■
   c) 5)2365
   d) 25)6254

2. Estimate to show that each answer in Question 1 is reasonable.

3. Calculate. Use mental math. Describe how you calculated one answer.
   a) 14.5 × 10 = ■
   b) 14 ÷ 10 = ■
   c) 123.2 × 100 = ■
   d) 12.3 ÷ 10 = ■
   e) 1.5 × 1000 = ■
   f) 15.75 ÷ 10 = ■

4. Each group of boxes of blueberries has a total mass of 10 kg. The boxes in each group have equal masses. Estimate the mass in kilograms of each box of blueberries. Explain what you did.

a)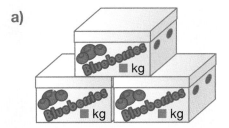

b)

c)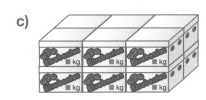

CHAPTER 10

1 Estimating Quotients

You will need
- a decimal place value chart

Goal Estimate quotients when dividing decimal numbers.

A sculptor wants to create a sculpture from iron rods, each 5.9 m in length. She plans to form five of the rods into **regular polygons**.

? How can you estimate the length of each side of a regular polygon?

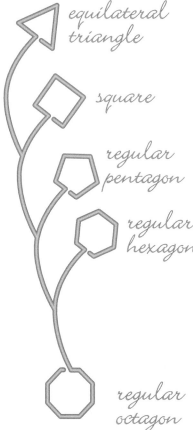

equilateral triangle

square

regular pentagon

regular hexagon

regular octagon

Maggie's Method

For the equilateral triangle, I need to divide 5.9 by 3.

To estimate 5.9 ÷ 3, I will round 5.9 to 6.

Khaled's Method

For the square, I need to calculate 5.9 ÷ 4.

To estimate 5.9 ÷ 4, I will round 5.9 to 6.0.

I can calculate 6.0 ÷ 4 by renaming 6 ones as 60 tenths.

Ones	Tenths
6	

→

Ones	Tenths
	60

294 NEL

Angele's Method

For the regular pentagon, I need to divide 5.9 by 5.

5.9 ÷ 5 means 5 × ■ = 5.9 or "5 times a decimal number equals 5.9."

I can estimate the value of ■ by multiplying these decimal numbers by 5.

5 × 1.0 = 5.0
5 × 1.1 = 5.5
5 × 1.2 = 6.0

I can use these answers to estimate 5.9 ÷ 5.

A. How can you tell without dividing whether the sides of each regular polygon will be greater than or less than 1 m?

B. Complete each student's estimation. Show your work.

C. How can you use your estimate of the length of each side of the equilateral triangle to estimate the length of each side of the regular hexagon?

D. How can your estimate of the length of each side of the square help you to estimate the length of each side of the regular octagon?

E. Show another way to estimate the length of each side of the regular hexagon and regular octagon.

Reflecting

1. Why do you think Khaled renamed 6.0 as 60 tenths rather than renaming 5.9 as 59 tenths?

2. Why do you think Angele stopped multiplying decimal numbers by 5 after she multiplied 1.2 by 5?

Checking

3. The sculptor also makes regular polygons from iron rods that are 11.8 m in length. Estimate the length of each side of three regular polygons. Show your work.

Practising

4. Sandy cut a large sheet of paper into 14 congruent rectangles. He pasted a plastic flag into each rectangle. Estimate the length and width of each rectangle. Show your work.

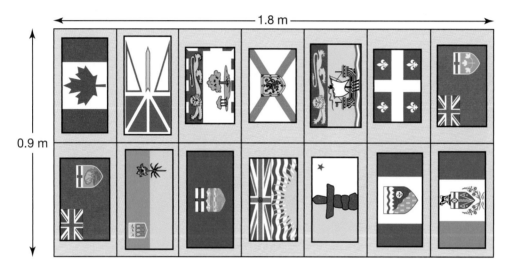

5. Estimate each quotient. Show your work.
   a) 1.7 ÷ 3 = ■
   b) 12.4 ÷ 6 = ■
   c) 15.7 ÷ 3 = ■
   d) 9.5 ÷ 2 = ■

6. You divide 8.4 by a one-digit whole number divisor and the answer is just less than 1. What is the divisor?

7. a) Estimate 12.7 ÷ 4. Show your work.
   b) Show how to use your answer in Part a) to estimate 12.7 ÷ 8.

8. Create a problem that you might solve by dividing 21.3 by 6.

9. When you divide a decimal number by a whole number your estimate is 3.4. Identify three possible pairs of numbers you might have been dividing.

Math Game

Estimate the Range

You will need
- a deck of 40 number cards (4 each of the digits from 0–9)
- game board
- a calculator
- counters

Number of players: 2 to 4

How to play: Estimate the quotient when dividing a decimal number by a one-digit number.

Step 1   Shuffle the cards. Deal out 4 cards.

Step 2   One player uses the first 3 cards to form a decimal number between 10 and 100, and the 4th card to be the divisor. (Draw another card if the 4th card is 0.)

Step 3   All players estimate the quotient and place counters on the box on the game board that best describes the quotient.

Game Board

Greater than 50	30 to 50	10 to 30	Less than 10

Step 4   Calculate the quotient. Use a calculator if necessary.

Step 5   If your estimate is in the correct range, you score 1 point.
Continue estimating and dividing until one player reaches 10 points.

Emilio's Turn

I put a counter on "Less than 10" because 13 is less than 20 and 20 ÷ 2 = 10. The answer is 6.55, which is less than 10, so I get 1 point.

CHAPTER 10

2 Dividing Money

You will need
- a calculator

Goal Solve problems by dividing money.

You and your friends decide to buy some DVDs to share.

? **How can you determine the cost for each person?**

A. Choose a DVD that you and one friend might want to buy. Determine the cost for each person. Use a calculator. Round your answer to two decimal places.

B. Use estimation to show that your answer in Part A is reasonable. Explain what you did.

C. Use multiplication to check your answer in Part A. Show what you did.

D. Choose another DVD that you and several friends might want to buy. Determine the cost for each person. Use a calculator. Explain how you know that your answer is reasonable.

E. Choose an item that you and some friends might want to buy. Use an actual price you found in a store, in an advertising flyer, or on a Web site. Determine the cost for each person. Explain how you know that your answer is reasonable.

Reflecting

1. Why did you not get the exact price of the item when you multiplied in Part C?

2. Why does it make sense to round each answer shown on the calculator to two decimal places?

Mental Math

Adding Decimals by Renaming

You can make the addition of decimals easier by renaming one or more of the numbers as a sum.

$$2.5 + 2.6 = 2.5 + 2.5 + 0.1$$
$$= 2.5 + 2.5 + 0.1$$
$$= 5.0 + 0.1$$
$$= 5.1$$

A. How did renaming 2.6 as 2.5 + 0.1 make the addition easier?

Try These

1. Calculate by renaming one or more numbers as a sum.
   a) 2.5 + 2.7
   b) 4.5 + 2.6
   c) 2.9 + 3.1
   d) 4.50 + 4.55
   e) 7.9 + 2.2
   f) 8.5 + 2.7 + 5.5

CHAPTER 10

3 Dividing Decimals by One-Digit Numbers

You will need
- grid paper

Goal Express quotients as decimal numbers to tenths.

Ayan's school district holds a track meet each year. In the 100 m hurdles for intermediate boys, the runners run 13.0 m to the first hurdle. They jump over 10 hurdles that are spaced equally. The last hurdle is 10.5 m from the finish line.

? What is the distance between each pair of hurdles?

Ayan's Solution

I drew a number line to represent this situation.

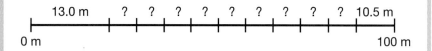

First I need to figure out the distance between the first hurdle and the tenth hurdle. I subtract 10.5 m and 13.0 m from 100 m and get 76.5 m.

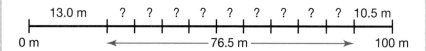

There are 9 equal spaces between the 10 hurdles. I need to calculate 76.5 ÷ 9 to determine the distance between each pair of hurdles. I estimate that the distance should be between 8 m and 9 m.

Step 1 I can't divide 7 tens by 9 to get a number of tens. So I divide 76 ones by 9 to get a number of ones.

```
      8
 9)7 6.5     76 ones ÷ 9 = 8 ones
   7 2       and 4 ones left over
     4
```

Step 2 I regroup the remainder of 4 ones as 40 tenths. Now I have 40 tenths + 5 tenths or 45 tenths to divide by 9.

```
      8
 9)7 6.5
   7 2
     4.5   ← 4.5 = 45 tenths
```

Step 3 I divide 45 tenths into nine equal groups of 5 tenths with 0 tenths remainder.

```
      8.5
 9)7 6.5
   7 2
     4.5     45 tenths ÷ 9
     4.5     = 5 tenths
       0
```

The distance between each pair of hurdles is 8.5 m.

Reflecting

1. Ayan estimated that the distance between each pair of hurdles was between 8 m and 9 m. Why is her estimate reasonable?
2. In Step 2, why do you think Ayan regrouped 4 ones as 40 tenths?
3. How would you multiply to check Ayan's answer?

Checking

4. In the 75 m hurdles for junior girls, the runners run 11.5 m to the first hurdle. They jump over eight hurdles that are spaced equally. The last hurdle is 11.0 m from the finish line.
   a) Draw a number line to represent the situation.
   b) Calculate the distance between each pair of hurdles. Show your work.
   c) Use multiplication and addition to check your answer.

Practising

5. Victoria has 5.0 kg of wild rice. She keeps one half for herself. She divides the remaining amount equally among five friends. How many kilograms of wild rice does each person get? Show your work.

6. Calculate.
   a) $12.5 \div 5 = \blacksquare$
   b) $6 \overline{)15.0}$
   c) $2.8 \div 4 = \blacksquare$
   d) $8 \overline{)13.6}$
   e) $23.7 \div 3 = \blacksquare$
   f) $7 \overline{)39.2}$

7. a) Choose one calculation in Question 6. Estimate to show that the answer is reasonable.
   b) Choose another calculation in Question 6. Multiply to check the answer.

8. Explain how you can use $3 \times 5 = 15$ to calculate 3×4.9.

9. William and five friends participated in *A Walking Tour of Canada*. One day, they walked a combined distance of 34.8 km.
   a) If each student walked the same distance, how far did each student walk?
   b) Use multiplication to check your answer.

10. Create and solve a problem in which you need to divide a decimal number dividend by a one-digit, whole-number divisor.

Curious Math

Dividing Magic Squares

You will need
- a calculator

In a magic square, the numbers in each row, each column, and each diagonal have the same sum. This total is called the **magic sum**.

9	6	3	16
4	15	10	5
14	1	8	11
7	12	13	2

1. What is the magic sum of this square?

2. Divide each number in the magic square by 2. Do you still have a magic square? If so, what is the magic sum?

3. Divide each number in the original magic square by 4. Do you still have a magic square? If so, what is the magic sum?

4. Divide each number in the original magic square by 5 or 8. Do you still have a magic square? If so, what is the magic sum?

5. What happens to the magic sum when each number in a magic square is divided by the same whole number?

CHAPTER 10

Frequently Asked Questions

Q. How do you estimate a quotient when dividing by a whole number?

A. For example, you can estimate 8.3 ÷ 4 by rounding 8.3 to 8 and then dividing 8 by 4.

8 ÷ 4 = 2, so 8.3 ÷ 4 is close to but greater than 2.

If you want to improve your estimate, you can rename 8.3 as 83 tenths. Because 84 is easier to divide by 4 than 83, you divide 84 tenths by 4 to estimate.

84 tenths ÷ 4 = 21 tenths or 2.1, so 8.3 ÷ 4 is close to but less than 2.1.

You might want to round up and round down to get a range of estimates. For example, you can estimate 43.5 ÷ 8 by both rounding 43.5 down to 40 and up to 48. Each of these numbers is easy to divide by 8.

40 ÷ 8 = 5 and 48 ÷ 8 = 6. So 43.5 ÷ 8 is between 5 and 6.

Q. How can you divide a decimal number by a one-digit whole number?

A. You can use pencil and paper. For example, 18.4 ÷ 4 can be calculated in several steps.

4 4)18.4 <u>16</u> 2	18 ÷ 4 = 4 ones, remainder 2 ones
4 4)18.4 <u>16</u> 2.4 24 tenths	Rename 2 ones as 20 tenths 20 tenths + 4 tenths = 24 tenths
4.6 4)18.4 <u>16</u> 2.4 <u>2.4</u> 24 tenths ÷ 4 = 6 tenths 0 or 0.6	Divide 24 tenths by 4.

CHAPTER 10

Mid-Chapter Review

LESSON 1

1. Danielle estimated each quotient. Explain what she might have done to estimate.
   a) 1.76 ÷ 3 is about 6 tenths.
   b) 4.98 ÷ 5 is just less than 1.
   c) 12.9 ÷ 2 is between 6 and 7.
   d) 15.89 ÷ 4 is close to 4.

2. Estimate. Show your work for one estimate.
   a) 2.9 ÷ 3   b) 11.1 ÷ 4   c) 7)4.8   d) 9)17.9

3. In Toronto, about 6.0 cm of snow usually falls in April. If the same amount of snow falls every week, estimate the weekly amount of snow for April. Show your work.

LESSON 2

4. One block of mild cheddar cheese costs $20.99. The block is divided into 5 equal pieces.
   a) Determine the cost of each piece of cheese. Use a calculator.
   b) Explain how you know your answer in Part a) is reasonable.

5. An MP3 player costs $89.99. It is on sale for half price.
   a) Estimate the sale price.
   b) Determine the sale price. Use a calculator.

LESSON 3

6. Calculate. Show your work.
   a) 12.4 ÷ 4   b) 24.5 ÷ 5   c) 7)6.3   d) 6)40.8

7. The total mass of a snowy owl and its six chicks is 3.1 kg. The mass of the owl is 1.9 kg. All the chicks have the same mass. What is the mass of each chick?

8. 8.1 L of pasta sauce is poured equally into three freezer bags. Each litre costs $2. What is the cost of each bag?

4. Dividing by 10, 100, 1000, and 10 000

You will need
- a decimal place value chart
- a calculator

Goal: Divide whole and decimal numbers by 10, 100, 1000, and 10 000 using mental math.

A souvenir shop sells models of large objects seen in Canada. The height of each model is close to 0.5 m. The height of a model is 10, 100, 1000, or 10 000 times smaller than the height of the actual object.

❓ **What is the height of each model?**

553 m

52.7 m

4.3 m

Li Ming's Method

I want to determine the height of a model of the CN Tower that is 10 times smaller than the actual tower.

I will divide 553 by 10. To divide 553 by 10, I represent 553 on a place value chart.

5970 m

Hundreds	Tens	Ones	• Tenths	Hundredths	Thousandths
5	5	3			

Each time I move the digits one place value to the right, each value become 10 times smaller.

Hundreds	Tens	Ones	• Tenths	Hundredths	Thousandths
	5	5	3		

553 ÷ 10 = 55.3

The height of a model that is 10 times smaller would be 55.3 m, which is too tall.

I divide 55.3 by 10 and 553 by 100 by moving the digits another place value to the right. The height of a model that is 100 times smaller is 5.53 m, which is still too tall.

Hundreds	Tens	Ones	Tenths	Hundredths	Thousandths
		5	5	3	

55.3 ÷ 10 = 5.53; 553 ÷ 100 = 5.53

If I divide 5.53 by 10 and 553 by 1000, I can determine the height of a model that is 1000 times smaller.

A. Complete Li Ming's method to determine the height of a model that is 1000 times smaller than the CN Tower.

B. Explain how you know your answer in Part A is reasonable.

C. Show how you can use multiplication to check your answer in Part A.

D. Use Li Ming's method to determine the height of the other models. Show your work.

E. Explain how you know that each answer in Part D is reasonable.

Reflecting

1. How did Li Ming know that the first two models of the CN Tower were too tall?

2. a) Why can you divide a number by 100 by dividing the number by 10 two times?
   b) Why can you divide a number by 10 000 by dividing the number by 10 four times?

3. Use mental math to divide 1350 by 10, 100, 1000, and 10 000. Explain what you did.

Checking

4. A souvenir shop sells each of these models.

i) A model that is 100 times smaller than the 62.5 m long hockey stick in Duncan, BC 	ii) A model that is 1000 times smaller than the 191 m high Calgary Tower in Calgary, AB 	iii) A model that is 10 000 times smaller than the 4020 m Waddington Mountain in the Coast Mountains, BC 

   a) Calculate the height of each model. Use mental math.
   b) Use multiplication to check each answer.

Practising

5. Calculate. Use mental math.
   a) 12.5 ÷ 10 = ■
   b) 10)¯103.5
   c) 2.8 ÷ 100 = ■
   d) 1000)¯77
   e) 960 ÷ 10 000 = ■
   f) 10 000)¯3210

6. Jonathan entered 234 into his calculator. He got this answer when he divided by a number. Which number did he divide by?

7. Pauline walked 113 m in 100 steps. What is the length of each of her steps?

8. The people of Smalltown spend $9840 each week on dog food. The town has 10 000 dogs. How much does a typical dog owner spend on dog food?

9. 48.3 L of pasta sauce is poured equally into 100 freezer bags. What amount of sauce is in each bag?

10. a) Calculate each pair.
    i)   0.1 × 123 = ■   123 ÷ 10 = ■
    ii) 0.01 × 123 = ■   123 ÷ 100 = ■
    iii) 0.001 × 123 = ■   123 ÷ 1000 = ■
    b) How are the two calculations in each pair related?
    c) Explain how you can multiply by a decimal to calculate 455 ÷ 1000.

Math Game

Calculate the Least Number

You will need
- a calculator
- a deck of 40 number cards (4 each of the digits from 0–9)
- a spinner

Number of players: 2 or more

How to play: Multiply or divide a four-digit whole number to determine the product or quotient.

Step 1   Shuffle the cards.
          Deal four cards to each player.

Step 2   Each player arranges the four cards to form a four-digit whole number.

Step 3   One player spins the spinner. Each player uses the operation spun to calculate the product or quotient.

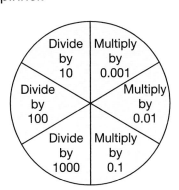

Step 4   The player with the least product or quotient scores 1 point.
          The first person to reach 10 points wins.

Tom's Turn

I used my four cards to create 2257.

I spun, "Divide by 1000."

$\boxed{2}\ \boxed{2}\ \boxed{5}\ \boxed{7} \div 1000 = 2.257$

My answer was the least quotient, so I score 1 point.

CHAPTER 10

5 Solving Problems by Working Backward

Goal Use a working-backward strategy to solve problems.

James's family is shipping a package that has a CD player and four speakers. The CD player has a mass of 4.5 kg. The four speakers are equal in mass. The total mass of the package is 7.7 kg.

? What is the mass of each speaker?

James's Solution

Understand

I need to determine the mass of each speaker. I know the total mass of the package and the mass of the CD player.

Make a Plan

I'll draw a diagram to represent the problem.

The diagram shows 4 speakers added to the mass of 4.5 kg. The total mass is 7.7 kg.

I can work backward to estimate and calculate the mass of one speaker.

Carry Out the Plan

I estimate that the mass of four speakers is about 3 kg, so the mass of one speaker is less than 1 kg.

Step 1
I subtract the mass of the CD player from the total mass.

The mass of the four speakers is 3.2 kg.

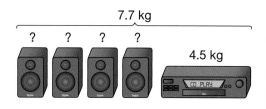

Step 2

I divide the mass of the four speakers by four to determine the mass of one speaker.

The mass of one speaker is 0.8 kg.

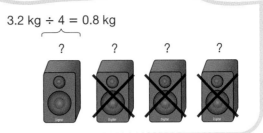

$3.2 \text{ kg} \div 4 = 0.8 \text{ kg}$

Reflecting

1. Explain how James might have estimated the mass of each speaker.

2. How can you use James's original diagram to help you check his answer?

3. Could James have divided first before subtracting? Explain.

Checking

4. Cathy measured the mass of six identical phones and an answering machine. The mass of the answering machine is 0.6 kg. The total mass of the phones and answering machine is 3.6 kg. What is the mass of each phone?
   a) Draw a diagram to represent this problem.
   b) Use the diagram to help you calculate the mass of one phone.

Practising

5. Thomas pours four identical jugs of cranberry juice and two bottles of ginger ale into a punch bowl. The total volume of the mixture is 9.0 L. Each bottle of ginger ale contains 1.5 L. How much does each jug of cranberry juice contain? Show your work.

6. Melissa thought of a number. She multiplied it by 2. She subtracted 2 from the product. The result was 3.6. What number did she start with? Show your work.

7. Create your own working backward problem. Give your problem to another student to solve.

CHAPTER 10

Skills Bank

1. To estimate 7.9 ÷ 4, Emily rounded 7.9 to 8.0. Show how to complete her estimate.

2. Estimate. Explain how you estimated one answer.
   a) 12.75 ÷ 6   b) 4.6 ÷ 8   c) 6)16.56   d) 7)0.85

3. This rectangle has an area of 20.7 m². Estimate its width.

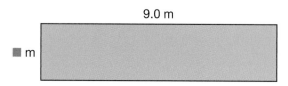

4. When Eric calculated 17.45 ÷ 3 on his calculator, he got this result.

   5.8166666 7

   Is his answer reasonable? Explain.

5. Brianna and her two friends bought a $26.99 music CD to share.
   a) Estimate the cost of each person's share.
   b) Determine the cost of each person's share. Use your calculator.

6. Determine each money amount. Use your calculator.
   a) $1.75 divided five ways
   b) $1.58 divided six ways
   c) $40.57 divided eight ways

7. A package of six small Métis flags cost $12.99.
   a) Determine the price of one flag. Use your calculator.
   b) Estimate to show that your answer is reasonable.

8. Calculate.
   a) 18.9 ÷ 9
   b) 5)42.5
   c) 7.6 ÷ 4
   d) 2)18.4
   e) 2.0 ÷ 5
   f) 6)13.2

9. a) Estimate to check one quotient in Question 8. Show what you did.
   b) Use multiplication to check another quotient in Question 8.

10. The perimeter of this regular hexagon is 9.0 m. What is the length of each side?

11. A rectangular park measuring 4.0 km by 4.6 km is divided into 8 equal sections. What is the area of each section?

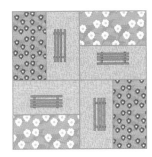

12. Michelle and her friends run on treadmills at their community centre. They recorded the total times and distances they ran.

Runner	Distance (km)	Time (h)
Michelle	18.6	6 h
Morgan	14.5	5 h
Kim	26.0	4 h
Christina	7.8	2 h

   a) How far did each person run each hour? Show your work.
   b) Use multiplication to check one answer from Part a). Show what you did.
   c) Use estimation to check another answer from Part a). Show what you did

13. Calculate. Use mental math.
    a) 52.5 ÷ 10 = ■
    b) 10)‾415
    c) 7.8 ÷ 100 = ■
    d) 100)‾47.7
    e) 155 ÷ 1000 = ■
    f) 10 000)‾4230

14. A 4.0 L container of apple juice is poured equally into 10 smaller containers. How much juice is in each container?

15. a) Calculate 45 ÷ 10.
    b) Explain how you can use your answer in Part a) to calculate 45 ÷ 100.

16. Shelby's bike wheels turn 10 000 times. If the bike travels 21 000 m, what is the distance around her wheels?

17. Laura subtracted 0.5 from her age, and divided that result by 3. The final answer was 4. How old is Laura?

18. The blue boxes are equal in mass. What is the mass of one blue box?

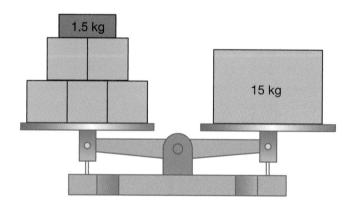

19. Erin received $40.00 for her birthday. She spent $7.50 for a movie ticket. Then she bought a bag of popcorn for herself and each of her three friends. Now she has $10.50 left. What is the cost of each bag of popcorn?

CHAPTER 10

Problem Bank

LESSON

1

1. Natalie used a calculator to divide a decimal number of the form ■.■■■ by 6. The answer was about 1.5. What decimal number might she have divided?

2. Choose numbers to place in ■.■■■ ÷ ■ so that when you use a calculator you get each quotient.
   a) Your quotient is a decimal just less than 1.
   b) Your quotient is close to, but not equal to, 0.1.
   c) Your quotient is close to, but not equal to, 0.5.
   d) Your quotient is close to, but not equal to, 4.5.

2

3. Four students share $10 equally. Each student spent half of the money he or she had. Determine how much money each student had left. Use your calculator.

3

4. The distances between the windows are equal.
   a) Estimate the distance between each pair of windows.
   b) Calculate the distance. Show that your estimate is reasonable.

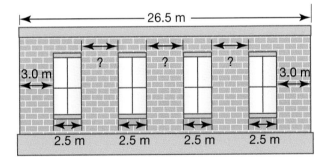

4

5. Aaron divides each of four different numbers by 10 000. Each time the digit representing the decimal one thousandths is 5. What might the numbers be?

5

6. Vanessa chose a number.
   She subtracted a number from her chosen number.
   She divided the difference by another number.
   She ended up with 6.5.
   What numbers might she have used?
   Show your steps.

CHAPTER 10

Frequently Asked Questions

Q. How do you divide a whole or decimal number by 10, 100, 1000, or 10 000?

A. Use place values. The digits in the result are the same as the digits you start with. Their place value just becomes 10 times, 100 times, 1000 times, or 10 000 times less. For example, to calculate 1560 ÷ 10, make the place value of each digit 10 times less.

Thousands	Hundreds	Tens	Ones	Tenths	Hundredths	Thousandths
1	5	6				

1 thousand becomes 1 hundred. 5 hundreds become 5 tens. 6 tens become 6 ones.

Thousands	Hundreds	Tens	Ones	Tenths	Hundredths	Thousandths
	1	5	6			

1560 ÷ 10 = 156

To divide a number by 100, divide by 10 twice.
 1560 ÷ 10 = 156
 156 ÷ 10 = 15.6

Thousands	Hundreds	Tens	Ones	Tenths	Hundredths	Thousandths
		1	5	6		

So, 1560 ÷ 100 = 15.6.

To divide a number by 1000, divide by 10 three times.
 1560 ÷ 10 = 156
 156 ÷ 10 = 15.6
 15.6 ÷ 10 = 1.56

So, 1560 ÷ 1000 = 1.56.

To divide a number by 10 000, divide by 10 four times.
 1.56 ÷ 10 = 0.156
So, 1560 ÷ 10 000 = 0.156

CHAPTER 10
Chapter Review

LESSON

1

1. Estimate. Show your work for one estimate.
   a) 22.9 ÷ 7   b) 8.1 ÷ 4   c) 7)4.8   d) 9)4.4

2. These signs are spaced equally along the road. Estimate the distance between each pair of signs.

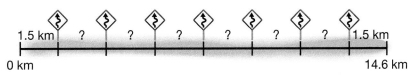

2

3. Five friends raised $36.96 for a charity. About how much money did each friend raise?

4. Kelly wants to buy a pair of FRS radios that are on sale for half price. She also wants to use a $20.00 gift certificate.
   a) Determine what she will pay for the radios. Use a calculator.
   b) Estimate to show that your calculation is reasonable.

3

5. Calculate. Show your work.
   a) 22.5 ÷ 5   b) 28.0 ÷ 8   c) 9)53.1   d) 6)1.2

6. Brooke cut a 5.6 m roll of exercise elastic into four equal pieces.
   a) Calculate the length of each piece.
   b) Use multiplication to check your calculation.

4

7. Calculate. Use mental math.
   a) 12.5 ÷ 10 = ■
   b) 10)21.5
   c) 34.8 ÷ 100 = ■
   d) 100)77.7
   e) 955 ÷ 1000 = ■
   f) 10 000)7230

8. Use multiplication to check two of your answers in Question 7.

9. The total mass of a bag of 100 clothespins is 920 g. What is the mass of each clothespin?

5

10. Steven's grandmother gives equal amounts of money to him and his two brothers. The three brothers each spend $10.00 on a movie. Now each has $2.50 left. How much money in total did their grandmother give them?

CHAPTER 10

Chapter Task

Judging the Fairness of a Game

Jorge and Denise are playing a game.

Each turn, Jorge rolls a die twice and records each number.

Denise divides the first number by the second number.

Jorge scores 1 point if the quotient is a whole number or has one decimal place.

Denise scores 1 point for any other quotient.

Denise scores 1 point because the quotient has more than 1 decimal place.

? How can you use the results of an experiment to determine if this game is fair?

A. Play 20 turns of the game. Tally the results.

Dice roll	Quotient	Jorge scores	Denise scores
1, 4	0.25		✓

B. Use the results of all students' experiments to make a classroom chart or tally chart.

C. Use the results of Part B to decide whether the game is fair or unfair. Explain what you did.

D. If the game is unfair, then use the results to explain how to change the scoring rules to make the game fairer.

E. Play the game with a classmate to decide if your scoring rules make the game fairer.

Task Checklist

☑ Did you check the reasonableness of quotients?

☑ Did you explain your thinking?

CHAPTER 11

3-D Geometry and 3-D Measurement

Goals

You will be able to
- **determine the surface area of polyhedrons**
- **estimate, measure, and calculate the volume of triangular prisms**
- **create views and isometric sketches of structures**
- **create structures from views and sketches**

Making nets

CHAPTER 11

Getting Started

Solving Net Puzzles

You will need
- scissors
- tape
- 3-D model set
- measuring cups
- water

I am made up of six congruent squares attached by their sides to form a T shape.

What am I?

I have a square in the middle with an isosceles triangle attached to each side. The four triangles are congruent.

What am I?

I have six congruent rectangles side by side attached to each other along their lengths. A pair of hexagons are attached to opposite ends of the fourth rectangle.

What am I?

? Can you make your own net puzzle?

A. Sketch nets of 3-D shapes from the puzzles.

B. What shapes might these nets fold into? Name them.

C. Identify objects in your classroom with shapes that match these nets.

D. Choose a different object from your classroom that matches a polyhedron in the 3-D model set. Draw its net.

E. Check that the net represents the polyhedron of the 3-D object by cutting and folding the net.

F. Make your own puzzle for the net and trade with a partner to solve.

Do You Remember?

1. Calculate the area of each polygon.

   a)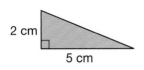

   c)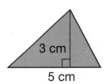

   b)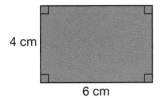

   d)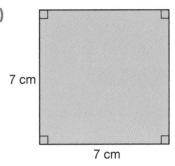

2. Determine the volume of each cube structure.

   a) 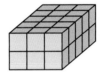 b)   c) 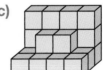   1 cm³

3. Calculate the volume of each prism.

   a) 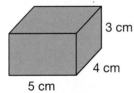 b) 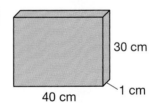 c)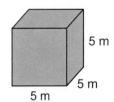

4. a) How many faces does a cube have?
   b) How many faces meet at each vertex of a cube?

CHAPTER 11

1 Visualizing and Constructing Polyhedrons

You will need
- polygons
- scissors
- tape

Goal Visualize and build polyhedrons from 2-D nets.

Emilio got a construction set for his birthday. He wants to know which pieces would fit together to make **polyhedrons**.

? **What polyhedrons can be made with these polygons?**

A. Create a net that can be folded into a polyhedron. Use congruent polygons only.

B. Fold to check. If the net works, sketch it and record what polyhedron it folds into.

C. Repeat Parts A and B with another set of congruent polygons.

D. Create a different net that can be folded into a polyhedron. Use any of the polygons.

E. Fold to check. If the net works, sketch it and tell what polyhedron it folds into.

F. Repeat Parts E and F as many more times as you can.

polyhedron
A 3-D shape with polygons as faces. Prisms and pyramids are two kinds of polyhedrons.

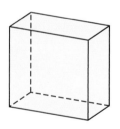

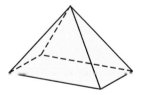

Reflecting

1. a) What did you notice about the nets of all of the pyramids you made?
   b) What did you notice about the nets of all of the prisms you made?

2. Why could you make more nets in Part D than Part A?

3. Which attributes of the polygons did you use to decide which polygons would fit together in nets?

Mental Imagery

Drawing Faces of Polyhedrons

You will need
- dot paper

This drawing on dot paper shows three faces of a polyhedron.

This drawing shows how to draw the yellow face.

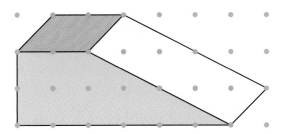
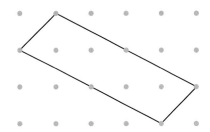

A. Draw the blue and green faces on dot paper.

B. Visualize and draw the three hidden faces.

Try These

1. Draw all of the faces of each polyhedron on dot paper.

   a)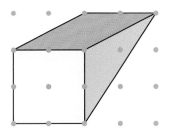

   b)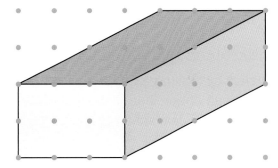

2. Draw a picture of a solid on dot paper. Have another student draw all the faces.

CHAPTER 11

Surface Area of Polyhedrons

You will need
- centimetre grid paper
- scissors
- a calculator

Goal Determine the surface area of triangular and rectangular prisms.

A chocolate shop is looking at designs for a new box. Each box holds the same number of chocolates. They have decided to hand-paint the boxes to decorate them. They want to pick the box that will need the least amount of paint.

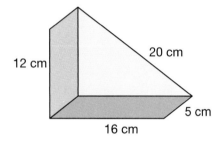

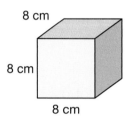

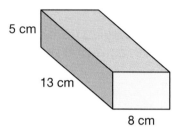

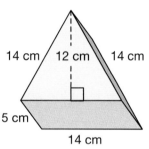

? **Which box requires the least amount of paint?**

Chandra's Calculation

I need to calculate the **surface area** of each box.

I'll start by making a net of one of the boxes.
Then I can check that I have included all the faces of the prisms in my calculation.

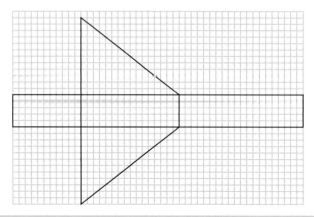

surface area
The surface area of a polyhedron is the total area of all of the faces, or surfaces, of that polyhedron.
For example, the surface area of this cube is 24 cm² because there are 6 faces and each face has an area of 4 cm².

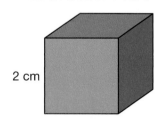

324

A. Draw each face of Chandra's first prism on grid paper. Cut out the faces and assemble them in a net. Fold your net to check.

B. Determine the area of each face and the surface area of the prism.

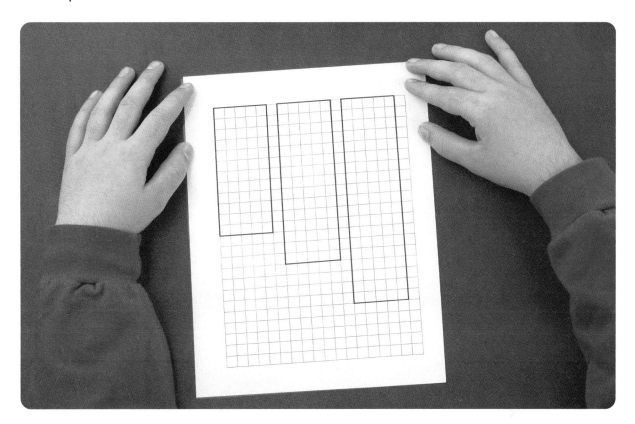

C. Determine the surface area of the other boxes.

D. Which box has the least surface area?

Reflecting

1. Which methods did you use to calculate the surface area of the prisms?
2. Could you use the same method for all of the prisms? Why or why not?
3. Does a polyhedron with more faces always have more surface area? Explain.

CHAPTER 11

3 Volume of Rectangular and Triangular Prisms

You will need
- base ten blocks

Goal Calculate the volume of rectangular and triangular prisms.

Maggie and Kurt baked a small cake and are going to share it fairly. The cake has a rectangular **base** 15 cm long by 10 cm wide. The height of the cake is 5 cm.

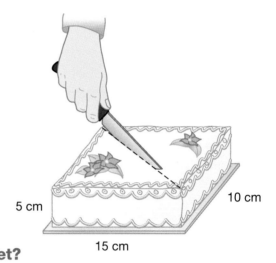

? How much cake does each person get?

Maggie's Strategy

I'll model the cake with layers of base ten blocks.

Each layer will be a rectangular prism 1 cm high.

I'll calculate the volume of the whole cake by multiplying the volume of each layer by the number of layers. Then I'll divide by 2 to calculate half.

I'll start by calculating the volume of the first layer.

The base of the prism is a rectangle 15 cm long and 10 cm wide, so the number of small cubes in the first layer is $15 \times 10 = 150$.

Each base ten cube has a volume of 1 cm³, so one layer of the cake has a volume of 150 cm³.

base
The base of a prism or pyramid is the face that determines the name of the prism.

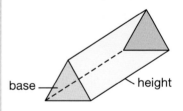

This is a triangular prism (also called a triangle-based prism) because its base is a triangle.

326

Kurt's Strategy

I imagine the cake is made up of layers 1 cm high.

Then I imagine cutting the cake into two congruent triangular prisms.

The volume of the first layer of the triangular prism is half the volume of the first layer of the cake.

To calculate the volume of the whole triangular prism, I'll calculate the volume of each layer of the prism, and then multiply by the number of 1 cm layers.

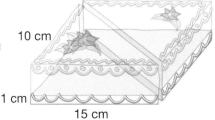

A. How many layers does Maggie need to complete her model? How do you know?

B. Complete Maggie's solution to calculate the volume of the whole cake.

C. What is the volume of Maggie's half?

D. Complete Kurt's strategy to calculate the volume of the triangular prism. What is the volume of his half?

Reflecting

1. Compare Maggie's and Kurt's methods. What do they have in common?

2. a) Write a rule for the volume of a rectangular prism in terms of the prism's length, width, and height.
   b) Does this rule apply to each layer of Maggie's cake? Explain.

3. a) Why can you also write the rule for a rectangular prism as volume = area of base × height?
   b) Does this rewritten rule also apply to Kurt's triangular prism? Explain.

Checking

4. a) Calculate the volume of the triangular prism.

   b) The slice is $\frac{1}{2}$ of a rectangular cake. What was the volume of the original cake?

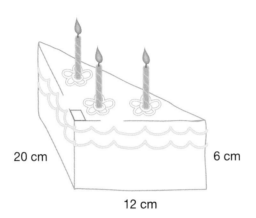

Practising

5. A foam company sells foam in different sizes and shapes. Brian needs a triangular prism of foam with a volume of 60 m³. Which rectangular prism should he choose to work with? Explain your choice.

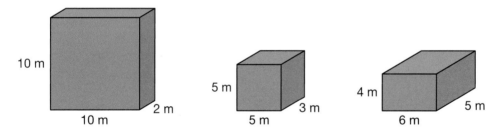

6. Calculate the volume of each triangular prism.

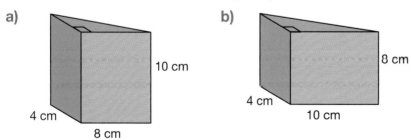

7. The volume of a triangular prism is 100 cm³. Describe 3 sets of possible dimensions for the prism.

Curious Math

Cross-Sections

You will need
- dental floss
- modelling clay

Ayan's Cross-Section

I made a triangular prism with modelling clay. The base is an isosceles triangle.

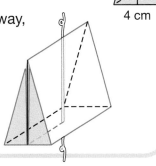

7 cm
6 cm
4 cm

If I cut this triangular prism this way, I get a rectangular face as my cross-section.

1 Make a triangular prism like Ayan's. Cut the prism to create cross-sections of different sizes and shapes. Record your cross-sections.

2 Measure the area of your cross-sections. How can you cut the prism to get a smaller area on the cross-section?

3 Repeat your experiment with a different prism.

CHAPTER 11

4 Solve Problems by Making a Model

You will need
- scissors
- a ruler
- centimetre cubes

Goal Make models to solve problems.

A brick of modelling clay is sold in a rectangular prism. The brick has a volume of 24 cm³ and it requires 56 cm² of shrink wrap to package it.

? How can you determine the dimensions of the prism?

Khaled's Plan

Understand
It is a rectangular prism of modelling clay.
The volume of the prism is 24 cm³.
The surface area of the prism is 56 cm².
I need to determine the dimensions of the prism.

Make a plan
I'll model the brick of clay using 24 centimetre cubes.
I'll make different rectangular prisms with the cubes and calculate the surface area of each one.

Carry Out the Plan

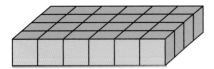

The surface area of this prism is 68 cm².

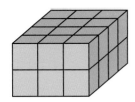

The surface area of this prism is 52 cm².

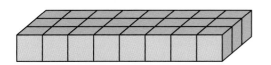

The surface area of this prism is 70 cm².

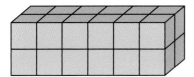

The surface area of this prism is 56 cm².

The dimensions of the rectangular prism of modelling clay are 2 cm × 2 cm × 6 cm.

Reflecting

1. Why did Khaled use 24 centimetre cubes?
2. How might Khaled have calculated the surface area of the prisms?
3. Was using a model a good way to solve the problem? Explain.

Checking

4. Erica made a rectangular prism using 36 centimetre cubes. The prism fits inside a 5 cm cube. What are the dimensions of the prism?

Practising

5. Which rectangular prism made with 8 centimetre cubes has the least surface area?
6. The top and sides of a stage are covered by 1 m × 1 m squares of construction paper. 20 squares cover the stage. What are the dimensions of the stage?

CHAPTER 11
Frequently Asked Questions

Q: How can you calculate the surface area of a polyhedron?

A. The surface area of a polyhedron is the sum of the areas of all of its faces.

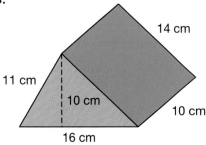

For example, to calculate the surface area of this triangular prism, first determine the area of the five faces.

There are two congruent triangular faces. The area of each triangle is half of 10 cm × 16 cm, which is 80 cm².

Determine the area of the rectangular faces:
10 cm × 11 cm = 110 cm²
10 cm × 14 cm = 140 cm²
10 cm × 16 cm = 160 cm²

The surface area of the triangular prism is:
80 cm² + 80 cm² + 110 cm² + 140 cm² + 160 cm² = 570 cm²

Q: How can you calculate the volume of prisms?

A: You can calculate the volume using the rule
volume = area of base × height.

For example, the area of the base of the green triangular prism is 80 cm² and its height is 10 cm.

So, the volume of the prism is 80 cm² × 10 cm = 800 cm³.

The volume of a rectangular prism can also be written as
volume = length × width × height.

So, the volume of this prism is 6 m × 2 m × 3 m = 36 m³.

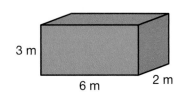

CHAPTER 11

Mid-Chapter Review

1. a) Name the polyhedron you can make with this net.
   b) Which side will join side BD when the net is folded?
   c) Which side will join side IJ when the net is folded?
   d) Which vertex will meet vertex A when the net is folded?

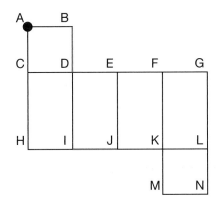

2. Calculate the surface area of each prism. Show your work.

   a)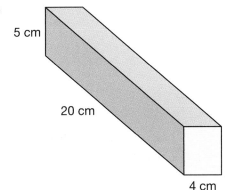

   b)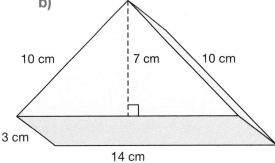

3. Calculate the volume of each prism in Question 2. Show your work.

4. a) Determine the dimensions of a rectangular prism with a volume of 18 cm³ and a surface area of 54 cm². Use a model.
   b) Determine the dimensions of a rectangular prism with the same volume but less surface area.
   c) Determine the dimensions of a rectangular prism with the same volume but greater surface area.

CHAPTER 11

5 Creating Isometric Sketches

You will need
- isometric dot paper
- a ruler
- pencil crayons
- linking cubes

Goal Sketch a polyhedron built from cubes.

Angele and Qi built a cube creature. Tom wants a copy of the creature, but they don't have enough cubes to build another one for him. They decide to tell him how to build the creature using his own cubes.

? How can they tell Tom how to build the creature?

isometric sketch
A 3-D view of an object that can be sketched on isometric dot paper. All equal lengths on the cubes are equal on the grid.

Angele's Instructions

Our cube creature is too big to sketch all at once, so I'll sketch each part separately. I'll start with the body.

Because the creature is made of cubes, I can use an **isometric sketch** to tell Tom how many cubes to use and where each cube goes.

The body is a rectangular prism.

Step 1 I line up the base of the prism with the dots on the paper.

Step 2 I start with the cube with the most faces showing. I sketch the faces I can see.

Step 3 I extend the edges to sketch the other cubes.

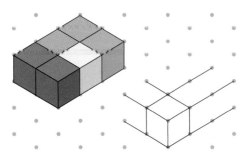
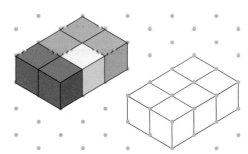

Qi's Drawing

I followed the same steps to sketch the body.

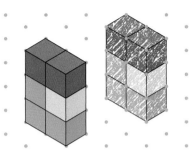

Communication Tip
The prefix "iso" means equal.
Isometric: equal distances between points
Isosceles: equal lengths

Reflecting

1. a) Why is it helpful to line up the base of the prism with dots on the paper before sketching?
   b) Could you start with a cube that doesn't have three faces showing? Explain.

2. a) Why are Angele's and Qi's sketches different?
   b) Tom says he can make a new isometric sketch of the body that is different from both of the drawings made so far. Explain how you would sketch it.

Checking

3. These five cubes form one of the cube creature's antennae. Make an isometric sketch.

Practising

4. a) Build something using up to 10 linking cubes.
   b) Sketch your cube structure on isometric dot paper.

5. a) Trade your cube structure from Question 4 with a classmate.
   b) Sketch your classmate's cube structure on isometric dot paper.
   c) Compare your sketch to your classmate's. Are they the same? Why or why not?

6. a) Model a letter of the alphabet using linking cubes.
   b) Sketch your model on isometric dot paper.

CHAPTER 11

6 Creating Cube Structures from Sketches

You will need
- isometric dot paper
- a ruler
- pencil crayons
- linking cubes

Goal Create cube structures based on an isometric sketch.

Tara and Tom are following Angele's isometric sketch to make the head of their cube creature. They can't decide how many cubes to use.

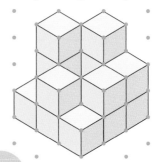

Tara's Construction

I made a cube structure that matches this sketch using 17 cubes.

Tom made a different one that matches the sketch using 14 cubes.

We need more information to build the cube creature's head correctly.

? **How can Angele make sure that they will build the head correctly?**

A. How many cubes are visible in Angele's isometric sketch?

B. Build a cube structure with 14 linking cubes that matches Angele's sketch.

C. Build a cube structure with 17 cubes that matches Angele's sketch.

D. Angele says the cube creature's head is made of 16 cubes. Make a cube structure with 16 cubes that matches Angele's sketch.

E. Make a second cube structure with 16 cubes that matches her sketch.

F. What additional instructions should Angele give to be sure other people build the head correctly?

Reflecting

1. How can both Tara's and Tom's models match Angele's sketch when they are made of different numbers of cubes?

2. What clues in the drawing did you use to help build the cube structure?

Checking

3. a) How many cubes are visible in this isometric sketch?
   b) Build a cube structure that matches the sketch. How many cubes did you use?
   c) Build another cube structure that also matches the sketch. How are your cube structures different?
   d) What additional information would you give with this isometric sketch to sketch your second structure?

Practising

4. a) Make a cube structure represented by this sketch.
   b) Sketch your cube structure so someone else would build it exactly as you did.

5. Build each cube structure.

   a)    b)    c)

6. All of these cube structures are made with six cubes. Which ones are the same?

   A.    B.    C.    D.

CHAPTER 11

7 Different Views of a Cube Structure

Goal Draw top, front, and side views of a cube structure.

You will need
- linking cubes
- grid paper

The instructions to Marc's 3-D jigsaw puzzle come with top, front, and side views of what the final structure should look like. He and Rebecca decide to make similar puzzles.

? How can you make top, front, and side views of a cube structure?

Rebecca's Drawings

I made the letter T using linking cubes.

I looked directly down at the structure to see the top view. The surface I saw is rectangular.

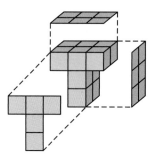

Next I brought my eye level with the structure to see the front view. The surface I saw is a T-shape.

Then I turned the cube so that I looked directly at the right-side view. The surface I saw is rectangular.

Marc's Drawings

I made a warehouse using linking cubes.

There is a change in depth from one layer of cubes to the next. I'll add a thick black line to show this change in depth when I draw the views on grid paper.

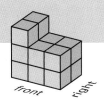

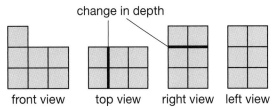

Reflecting

1. Why don't the different views of a cube structure always show the same number of cubes?

2. Why did Marc make side views from the left and right?

3. Can two cube structures have the same top view but different side views? Explain, using an example.

Checking

4. Build a model chair with 10 to 20 linking cubes.
   a) Draw the top view of your chair.
   b) Draw the front view of your chair.
   c) Draw the side view of your chair.

Practising

5. Make a cube structure that looks like this from the top and from the front.

6. a) Use up to 20 linking cubes to make an airplane that looks different from the top, front, and side.
   b) Draw its top, front, and side views.

7. Make a rectangular prism out of linking cubes. Draw its top, front, and side views.

8. What would the top, front, and side views of this prism look like? Explain how you know.

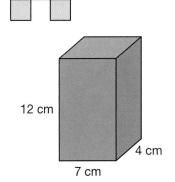

339

8 Creating Cube Structures from Different Views

You will need
- linking cubes

Goal
Make cube structures when given their top, front, and side views.

? How can you build a cube structure using their top, front, and side views?

Isabella built a model skyscraper using 16 linking cubes. She drew top, front, and side views of the model.

A. Make several different cube structures that match the top view.

B. Make several different cube structures that match the side view. Do any of your cube structures match both the top and the side views?

C. Make several different cube structures that match the front view. Do any of your cube structures match all three of Isabella's views?

top

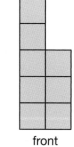
front

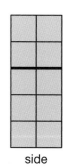
side

Reflecting

1. Is it possible to make more than one cube structure that matches Isabella's top, side, and front views? Explain.

2. How did you use the views to help you figure out what the model could look like?

Curious Math

Plane of Symmetry

You will need
- modelling clay
- dental floss
- a transparent mirror

A 3-D shape can have **mirror symmetry** just like a 2-D shape. Instead of a line of symmetry dividing a shape, a plane of symmetry divides the shape.

Farmers in Japan have succeeded in making cubic watermelons. They now have spherical, oval, and cubic watermelons.

mirror symmetry
The property of a shape such that it can be divided into two halves that are mirror images of each other

Make models of symmetrical watermelons with modelling clay. Make them into different shapes.

1. Identify a line of symmetry on a face of your cube model. Cut the model by aligning dental floss with your line and pulling the floss through the model. Are the halves mirror images? If so, you have just cut along a plane of symmetry for your model.

2. How could you cut an oval-shaped watermelon so that the cut is along a plane of symmetry?

3. Identify as many planes of symmetry as you can by locating the lines of symmetry on the faces of your models.

4. Why are there more planes of symmetry on an oval-shaped watermelon than on a cube watermelon?

CHAPTER 11

Skills Bank

LESSON

1
1. a) Sketch nets of three different pyramids using these polygons. You may use the polygons more than once. Name the polyhedrons.
   b) Sketch nets of three different prisms using these polygons. You may use the polygons more than once. Name the polyhedrons.

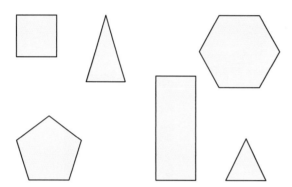

2
2. Determine the surface area of each prism.

   a)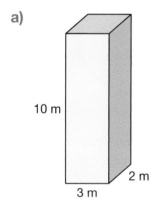

   b)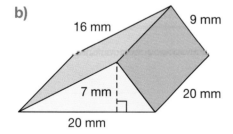

   c)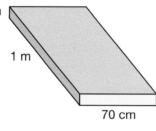

   d)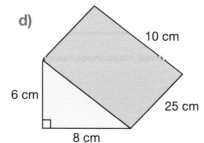

3. Determine the volume of each prism in Question 2.

4. The surface area of a rectangular prism is 96 cm².
   a) What are the dimensions of this prism if it is a cube? Explain.
   b) What is the volume of the prism?

5. a) Build a cube structure using 15 linking cubes.
   b) Make isometric sketches of it from two different views.

6. Trade your isometric sketches from Question 5 with a partner.
   a) Build your partner's cube structure.
   b) Did your cube structure match your partner's? Why or why not?

7. a) Draw the top view of this cube structure.
   b) Draw the front view.
   c) Draw a side view.

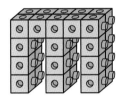

8. Build each cube structure using linking cubes.

   a)
   top view   front view   side view

   b)
   top view   front view   side view

   c)
   top view   front view   side view

CHAPTER 11

Problem Bank

1. Can these polygons make a net? Why or why not?

2. Calculate the surface area of each polyhedron.

   a)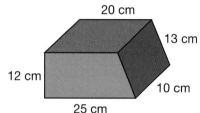

   b)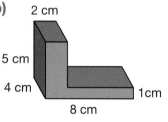

3. a) A rectangular prism has a volume of 2400 cm³. The base of the prism has an area of 300 cm². What is its height?
   b) What is the height of a triangular prism with the same volume and same area for the base?

4. Kristen made this isometric sketch.
   a) What is the least number of cubes that could be in this structure?
   b) Is there a greatest number of cubes? Explain.

   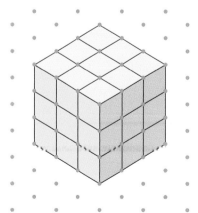

5. Travis drew the top view of a square-based prism made with centimetre cubes. Calculate a possible volume for the prism.

   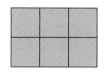

CHAPTER 11

Frequently Asked Questions

Q: How do you make an isometric sketch of a cube structure?

A: Line up the vertices of the cubes with the dots of the paper. Then start by sketching a cube with three faces showing. Extend the lines to sketch the cubes around it.
Continue sketching the cubes until the isometric sketch matches the cube structure.

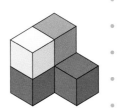 

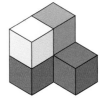 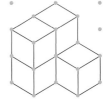

Q: Why is it possible to have different isometric sketches of the same cube structure?

A: You can make different isometric sketches of the same cube structure by sketching it from a different view. For example, three cubes can be sketched on isometric dot paper in several different ways. Each picture is a different view of the same three cubes.

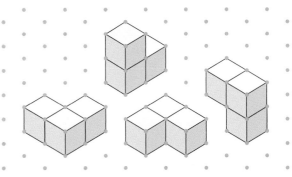

Q: Why do you need multiple views of a structure in order to build it properly?

A: Structures that are different can share the same top views but have different front or side views.
For example, these cube structures have the same top view, but their side views are different.

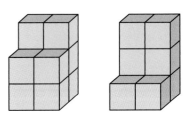

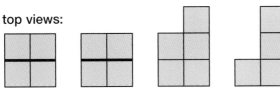

345

CHAPTER 11

Chapter Review

1. Describe the polygons and number of each you need to make nets for each polyhedron.
   a) a pyramid with a regular hexagon base
   b) a prism with a regular octagon base
   c) a pyramid with a rectangular base
   d) a prism with an isosceles triangle base

2. Caleb plans to decorate a block of wood as a gift. He wants to choose the one with the least surface area to decorate. Which one should he choose? Explain your answer.

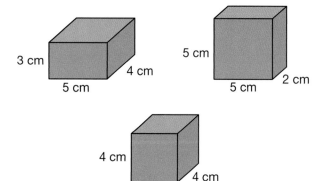

3. Which block from Question 2 has the greatest volume? Show your work.

4. Calculate the surface area and volume of each triangular prism.

a)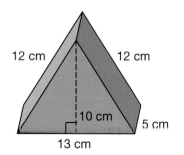

b)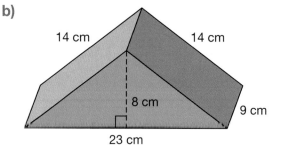

5. Crystal built a rectangular prism with linking cubes. Its surface area was 52 cm² and its volume was 24 cm³. Use cubes to build a cube structure with the same volume and surface area.

6. a) Use 20 linking cubes to build a cube structure.
   b) Make an isometric sketch of your model.
   c) Make an isometric sketch from a different view.

7. Jesse built a model snake using 10 linking cubes. He made an isometric sketch of his snake. Build his model using linking cubes.

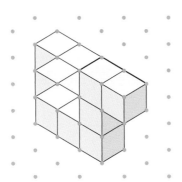

8. Jared built a model canoeist using 11 linking cubes. He made an isometric sketch of his model. Build his model using linking cubes.

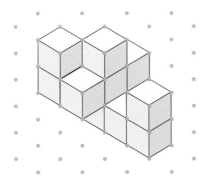

9. Pick a number from 0 to 9. Make a model of your number using linking cubes. Draw the top, front, and side views of your model.

10. Build a cube structure. Use up to 30 linking cubes.
    a) Draw the top, front, and side views of the structure.
    b) Work with a partner. Make your partner's structure using the top, front, and side views.
    c) Did the structure you built match your partner's? Why or why not?

CHAPTER 11

Chapter Task

Painting Bids

The owners of this building want the roof and walls painted. Paint-A-Wall is planning to bid on the project.

Paint-A-Wall can buy paint that costs $2.00 for every 1 m². It will also charge $1000 for labour costs and other expenses.

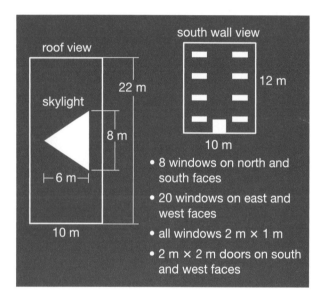

- 8 windows on north and south faces
- 20 windows on east and west faces
- all windows 2 m × 1 m
- 2 m × 2 m doors on south and west faces

? **How much should Paint-A-Wall bid?**

A. Create drawings to represent the other walls of the building.

B. Make a model of the building. Use a scale of 1 cm represents 1 m.

C. Determine the area that needs to be painted on each wall and the roof.

D. Determine the cost of painting the roof and side walls.

E. Add the other costs to the cost of the paint to determine the bid.

Task Checklist
- ✓ Did you use math language?
- ✓ Did you label your drawings and diagrams?
- ✓ Did you draw your diagrams accurately?

CHAPTERS 8–11

Cumulative Review

Cross-Strand Multiple Choice

1. Which is the area of this parallelogram?
   A. 8 cm²   C. 16 cm²
   B. 32 cm² D. 4 cm²

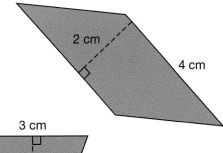

2. Which is the area of this triangle?
   A. 12 cm² C. 24 cm²
   B. 7 cm²   D. 6 cm²

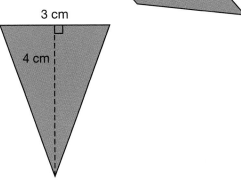

3. Which is the area of this triangle?
   A. 2 cm²   C. 12 cm²
   B. 6 cm²   D. 4 cm²

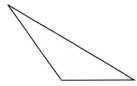

4. The area of this triangle is 36 cm². Which is the missing dimension?
   A. 4 cm²   C. 8 cm
   B. 2 cm   D. 6 cm

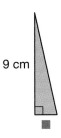

5. Ayan made a triangle poster with a base of 50 cm and a height of 20 cm. Tom made a rectangle poster with the same area and base as Ayan's poster. Which are the dimensions of Tom's poster?
   A. 50 cm by 40 cm   C. 50 cm by 20 cm
   B. 25 cm by 8 cm   D. 50 cm by 10 cm

6. The capacity of a thermos is advertised as 2.645 L. Which is the capacity in millilitres?
   A. 264.5 mL   B. 2645 mL   C. 264.5 mL   D. 26 450 mL

7. Finger nails grow 2.5 cm every year. How much do fingernails grow in seven years?
   A. 17.5 cm   B. 1.75 cm   C. 175 cm   D. 14.5 cm

8. Each bead in a craft set is 0.01 m long. Rebecca made a window decoration with 98 beads on a string. She left no spaces between the beads. Which is the length of all the beads on the string?
   A. 0.098 m   B. 9.8 m   C. 98 m   D. 0.98 m

9. Nine students shared 32.4 kg of sand equally for their science projects. Which is the amount for each student?
   A. 3.0 kg   B. 0.36 kg   C. 3.6 kg   D. 36 kg

10. James's class has 39.5 L of silver paint to sell at a fair. If they pour an equal amount of the paint into each of 100 jars, which is the amount for each jar?
    A. 395 L   B. 0.395 L   C. 39.5 L   D. 3.95 L

11. Emilio made this triangular prism gift box. Which is the surface area?
    A. 690 cm²   B. 420 cm²   C. 431 cm²   D. 504 cm²

12. Which is the volume of Emilio's box?
    A. 1008 cm³   B. 540 cm³   C. 504 cm³   D. 1080 cm³

13. Each of these cube structures is made with eight cubes. Which is not the same structure as the other three?

    A.    B.    C.    D.

14. Tara built a structure with centimetre cubes and drew these views. Which is the volume of Tara's structure?

    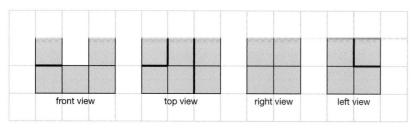

    A. 9 cm³   B. 8 cm³   C. 10 cm³   D. 11 cm³

Cross-Strand Investigation

The Peace Tower

Some students are entering a contest to win a trip to a tower of their choice. To enter, they must research data about the tower. Isabella chose the Peace Tower in Ottawa.

15. a) The area of the Canadian flag at the top of the Peace Tower is about 9.9 m². What is the area in square centimetres?
    b) The Peace Tower is 92.2 m tall. Isabella saw a model of the Peace Tower advertised on the Internet. The model is 100 times smaller than the Peace Tower. How tall is the model? Show how to multiply to find the height. Show how to divide to find the height.
    c) Isabella reseached the triangles on the copper roof. Each triangle is 20 m high and has a base of 8 m. What is the total area of the copper in the four triangles?
    d) A gargoyle decorates each corner near the top of the Peace Tower. Each gargoyle is 2.5 m long. What is the total length of the four gargoyles?
    e) Build a tower with 15 to 20 cubes. Put at least three cubes in the bottom layer.
    f) Make an isometric sketch of your tower.
    g) Draw top, front, and side views of your tower on centimetre grid paper.

16. a) Each gargoyle on the Peace Tower is 0.75 m high. How many centimetres high is each gargoyle?
    b) The Peace Tower replaced the Victoria Tower, which was destroyed in a fire in 1916. Isabella found a website that said the Victoria Tower was 54.865 m tall. What was its height to the nearest tenth of a metre?
    c) What is an appropriate unit to measure the width of the Peace Tower? Explain your choice.
    d) Isabella discovered that the Peace Tower is decorated with 368 stone carvings. What is the mean number of stone carvings for each of the four sides?
    e) Isabella wrote this equation ■ − 5 = 8 × 6. The missing number is the number of bells in the Carillon at the top of the Peace Tower. How many bells are there?

The Eiffel Tower

Li Ming choose the Eiffel Tower in Paris, France, built in 1889. She researched data about the Eiffel Tower platforms.

Platform	Number of steps to platform	Height of platform (m)	Area of platform (m²)
First	347	57.6	4200
Second	674	115.8	1400
Third	1710	276.1	350

17. a) Li Ming said the third platform is about five times as high as the first platform. Explain why her estimate is reasonable.
    b) About how many times as high as the first platform is the second? Justify your answer.
    c) Sketch a parallelogram and a triangle each with the same area as the first platform. Label the base and the height of each shape.
    d) Li Ming has a book about the Eiffel Tower. The front of the book is 20 cm by 18 cm. It is 2 cm thick. She made as small a cardboard box as possible for her book. How much cardboard is needed for the box without any overlapping?

18. a) The height of the Eiffel Tower can increase or decrease up to 15 cm because of the temperature. Li Ming read that the tower was 324.25 m high, including the 23.73 m antenna. Explain how to estimate the height of the tower without the antenna.
    b) Li Ming read that the names of 18 scientists are engraved on each of the four faces of the Eiffel Tower. She calculated the total number of names by thinking $4 \times 20 - 4 \times 2 = 80 - 8$, or 72. Explain her strategy.
    c) Li Ming calculated the number of steps from the first to the second platform by thinking $677 - 350 = 327$. Explain her strategy.
    d) Calculate the number of steps from the second to the third platform. Use mental math. Explain your strategy.

CHAPTER 12

Fractions, Decimals, Ratios, and Percents

Goals

You will be able to

- relate and compare fractions, decimals, and percents
- identify, model, and apply ratios in various situations
- represent relationships using unit rates
- solve problems using a guess and test strategy

Experimenting with a globe

CHAPTER 12

Getting Started

You will need
- counters
- grid paper
- pencil crayons

Filling a Pancake Order

Kurt's class is having a pancake breakfast. Each student ordered one type of pancake in advance. $\frac{1}{2}$ of the students ordered chocolate chip, $\frac{1}{6}$ of the students ordered blueberry, $\frac{1}{4}$ of the students ordered plain, and the rest ordered strawberry.

? **What fraction of the class ordered strawberry pancakes?**

A. Could there be 16 students in the class? Could there be 26 students in the class? How do you know?

B. What do the fractions of students who ordered each type of pancake tell you about the class size?

C. Determine a possible class size. Show your work.

D. For the class size you chose in Part C, determine the number of each kind of pancake they need to make.

E. What fraction of the class ordered strawberry pancakes?

Do You Remember?

1. What fraction of each whole is coloured?
   Name an equivalent fraction for each picture.

   a)   b)   c)

2. Match each number on the left with an equivalent representation on the right.

a)	0.4	A	
b)	0.2	B	
c)	$\frac{5}{9}$	C	four tenths
d)	$\frac{1}{4}$	D	
e)	$\frac{3}{2}$	E	
f)	0.5	F	
g)	$1\frac{1}{4}$	G	1.5

3. Compare. Write >, <, or =. Explain your choice.

   a) $\frac{7}{4}$ ■ $\frac{3}{4}$ c) $\frac{2}{5}$ ■ $\frac{4}{10}$ e) $\frac{11}{5}$ ■ $\frac{2}{5}$

   b) $\frac{4}{10}$ ■ $\frac{7}{10}$ d) $\frac{9}{10}$ ■ $1\frac{1}{10}$ f) $\frac{6}{12}$ ■ $\frac{1}{4}$

4. Write a decimal equivalent for each. Show your work.

   a) $\frac{4}{5}$ b) $\frac{7}{10}$ c) $3\frac{1}{2}$ d) $\frac{5}{2}$

5. Write an equivalent fraction for each.

   a) 0.73   b) 0.25   c) 0.6   d) 1.5

CHAPTER 12

1 Comparing and Ordering Fractions

You will need
- fraction strips
- number lines

Goal Compare and order fractions on number lines.

All families in Meadowcreek are supplied with containers that hold four bags of garbage each. The weekly limit for free pick-up is one container. Every bag over the limit costs $2.00.

Jorge is in charge of the family garbage.

? How much will Jorge's family pay for garbage pick-up each week?

Jorge's Solution

I'll place the number of containers for each week on a number line.

Since four bags represent one full container, I need to divide the line into fourths. A fraction strip will help.

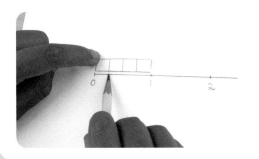

This Month's Garbage

Week	Number of filled containers
Week 1	1
Week 2	$\frac{3}{4}$
Week 3	$2\frac{1}{2}$
Week 4	$\frac{6}{4}$

A. How long should the number line be to show the number of containers for each of the four weeks?

B. Draw Jorge's number line. Label the number of containers for each week.

C. List the fractions in order from least to greatest.

356

D. In which weeks did Jorge's family fill more containers than in Week 1? Compare using an **inequality sign**.

E. How many bags over the limit did Jorge's family put out each week?

F. How much will Jorge's family pay for garbage pickup each week?

Reflecting

1. How did you decide where to put the **improper fraction** $\frac{6}{4}$?
2. Explain your strategy for ordering the numbers.

Checking

3. Jorge's family makes liquid fertilizer from their fruit and vegetable waste.
   a) Which fraction strip will help you draw a number line with a place for each number?
   b) Use an inequality sign to compare the amounts made in Week 1 and Week 2.
   c) List the amounts of fertilizer from least to greatest.

Week	Amount of fertilizer (buckets)
Week 1	$1\frac{1}{5}$
Week 2	$\frac{7}{5}$
Week 3	$\frac{3}{5}$
Week 4	$2\frac{2}{5}$

Practising

4. Compare. Write >, <, or =. Explain your strategy.
   a) $\frac{5}{4}$ ■ $\frac{6}{4}$
   b) $\frac{3}{2}$ ■ $3\frac{1}{2}$
   c) $1\frac{1}{2}$ ■ $1\frac{3}{4}$
   d) $\frac{2}{5}$ ■ $\frac{3}{10}$

5. Order each set of numbers from least to greatest.
   a) $\frac{3}{4}, \frac{4}{2}, 2\frac{1}{2}, \frac{7}{4}$
   b) $2, \frac{8}{10}, \frac{6}{5}, \frac{9}{5}, \frac{9}{10}, 1\frac{1}{10}$

6. The chart shows hours spent doing weekend chores.
   a) Which fraction strip will help you draw a number line with a place for each number in the chart? Explain your thinking.
   b) Use an inequality symbol to compare the number of hours Chris and Gabriel spent on chores.
   c) Order the fractions in the chart from least to greatest.

Name	Time to complete chores (in hours)
Mia	$\frac{5}{4}$
Chris	$\frac{9}{4}$
Katherine	$1\frac{1}{2}$
Andrea	$\frac{3}{4}$
Gabriel	$1\frac{3}{4}$

CHAPTER 12

2 Comparing Fractions with Unlike Denominators

You will need
- grid paper
- fraction strips
- a calculator

Goal Compare fractions when the denominators are different.

The 15 km Charity Walk-a-thon has
- a trail mix station every two thirds of a kilometre;
- a water station every three fourths of a kilometre; and
- a cooling station every three halves of a kilometre.

Ayan has reached the first water station, Mark is at the first cooling station, and Angele is at the second trail mix station.

? Who has walked the farthest?

Raven's Solution

I'll make a sketch of the first 2 km of the race on grid paper and mark the stations. I need to show halves, thirds, and fourths, so I want a whole that I can easily divide by 2, 3, or 4. I think a whole with 12 sections will work. 1 km will be represented by 12 squares.

A third of 12 squares is 4 squares, so $\frac{1}{3}$ is 4 squares past 0 and $\frac{2}{3}$ is another 4 squares past $\frac{1}{3}$.

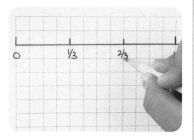

A. Draw a number line for 0 km to 2 km. Mark all of the thirds. Label the trail mix stations.

B. Mark the fractions for the cooling stations.

C. Mark the fractions for the water stations.

D. Mark the locations of the three students.

E. Who has walked the farthest?

Reflecting

1. How did you decide where to put the water and cooling stations?

2. Look at the positions of these fractions on the number line: $\frac{2}{2}$, $\frac{2}{3}$, and $\frac{2}{4}$. Which fraction is greatest? Why does that make sense?

3. How can you compare fractions when both the numerators and denominators are different? Use $\frac{3}{4}$ and $\frac{2}{3}$ as an example.

Checking

4. a) Write fractions for the locations of these stations:
   b) Order the fractions in Part a) from greatest to least.
   c) Explain the strategies you used to order the fractions.

- T, the fourth trail mix station
- W, the third water station
- C, the third cooling station

Practising

5. Compare. Write >, <, or =. Explain your strategy.
   a) $\frac{5}{6}$ ■ $\frac{1}{6}$ b) $\frac{2}{4}$ ■ $\frac{2}{5}$ c) $1\frac{1}{2}$ ■ $\frac{3}{4}$ d) $\frac{5}{2}$ ■ $3\frac{1}{2}$

6. For each pair of stations in the Charity Walk-a-thon, which is farther from the start of the race? How do you know?
   a) The second water station or the third trail mix station
   b) The third cooling station or the fourth water station
   c) The sixth trail mix station or the fourth cooling station

7. For each pair of chores, which one took longer to complete? Tell how you know.
   a) $\frac{4}{5}$ h doing laundry or $\frac{2}{5}$ h vacuuming
   b) $\frac{1}{3}$ h washing dishes or $\frac{1}{5}$ h drying dishes
   c) $\frac{1}{2}$ h collecting garbage or $\frac{3}{5}$ h cleaning the bathroom

8. Count by fourths from 0 to 4.

9. What is the greatest value you can use to make each true?
   a) $\frac{■}{5} < \frac{3}{4}$ b) $3\frac{2}{3} < 3\frac{4}{■}$ c) $4\frac{3}{8} > ■\frac{2}{3}$

CHAPTER 12

3 Fraction and Decimal Equivalents

You will need
- fraction strips
- number lines
- play money
- a calculator

 Goal Relate fractions to decimals and determine equivalents.

Tara wrote a story and removed the numbers to create a puzzle.

> Maggie needs $■ for a one-way bus fare to visit her Grandma. In her piggy bank she only has $■, which is ■ of a dollar. She asks her Mom for $■ to make up the difference. Another name for that amount is ■ of a dollar. If Maggie were to return home on the bus as well, she would need a total of ■ of a dollar.

These are the numbers from Tara's story:

$\frac{2}{5}$, 0.90, 0.50, $\frac{18}{10}$, 0.40, $\frac{1}{2}$

? **How can you make a story puzzle like Tara's?**

A. Copy the story and replace each ■ with one of the numbers listed so that the story makes sense.

B. How did you decide which number to use in each spot?

C. Write your own story that uses about five fractions and decimals. Make sure you use both fractions and decimals.

D. Create a puzzle for a classmate to solve by rewriting your story with the numbers taken out and listing them in a different order.

Reflecting

1. Explain how to write a fraction like $\frac{2}{5}$ as a decimal equivalent without using a calculator.

2. How do you write the decimal 0.25 as an equivalent fraction?

Mental Math

Using Factors to Multiply

Sometimes, you can use the factors of a number to multiply with mental math.

To calculate 4 × 75, list factors of 75.

4 × 25 × 3

100 × 3 = 300

A. What is another way you can use mental math to multiply 75 by 4?

Try These

1. Calculate.
   a) 4 × 15
   b) 8 × 25
   c) 12 × 15
   d) 12 × 25
   e) 4 × 75
   f) 12 × 75

2. How can your answer to Question 1e) be used to determine 4 × 7.5?

3. Explain how knowing 4 × 25 = 100 can help you calculate each product.
   a) 16 × 25
   b) 4 × 250
   c) 5 × 25

CHAPTER 12

4 Ratios

You will need
- counters

Goal
Identify and model ratios to describe situations.

Rodrigo is trying a painting technique he read about in an art book. He mixes different **ratios** of green and white paint to make four distinct tints.

He wants to order the shades of paint from lightest to darkest.

Tint	Number of containers of green paint	Number of containers of white paint
A	3	0
B	3	1
C	3	2
D	3	3

? What order should Rodrigo use for the tints?

Rodrigo's Solution

I will use counters to model the mixtures. I decided to write the ratios as green : white instead of white : green.

When I start, the ratio of green paint to white paint is 3 : 0.

A green : white = 3 : 0

B green : white = 3 : 1

C green : white = 3 : 2

D green : white = 3 : 3

The ratio with the most white paint compared to green represents the lightest green.

From lightest to darkest I should list D, C, B, A.

ratio
A comparison of two numbers or quantities measured in the same units.

If you mix juice using 1 can of concentrate and 3 cans of water, the ratio of concentrate to water is 1 : 3, or 1 to 3.

Communication Tip
For ratios, read the symbol : as "to."
Ratios can be written as **part** to **part** or **part** to **whole**.
In the juice example, concentrate to water, 1 : 3, is a part to part ratio
and concentrate to juice, 1 : 4, is a part to whole ratio.

Reflecting

1. Rodrigo wrote the green : white ratio for each mixture. What other part: part ratios might he have chosen to write? What part to whole ratios?

2. When you are working with a part to part ratio, how do you determine what the whole is?

3. What happened to the total amount of paint when Rodrigo added more and more white parts?

Checking

4. Rodrigo decided to make different shades of red. His first mixture used 3 cups of red paint to 2 cups of white paint.
   a) Write the ratio of the number of cups of red paint to the number of cups of white paint.
   b) Write the ratio of the number of cups of red paint to the total number of cups of paint.
   c) Write the ratio of the number of cups of white paint to the total number of cups of paint.

Practising

5. a) Write the ratio of the number of pucks to the number of sticks.
   b) Write the ratio of the number of sticks to the number of pucks.

6. a) Write the ratio of the number of DVDs to the number of cases.
   b) Write the ratio of the number of cases to the number of DVDs.

7. Jason is mixing lemonade from concentrate. The recipe reads, "Mix the concentrate and water in the ratio 1 to 4."
   a) Model the ratio using counters. Sketch your model.
   b) If Jason uses 1 cup of concentrate, how much lemonade will he make?
   c) How else can you describe this mixture using part to part or part to whole ratios? Include words and numbers for each ratio.

CHAPTER 12

5 Equivalent Ratios

You will need
- counters
- base ten blocks

Goal Determine equivalent ratios and use them to solve problems.

Akeem found a collection of 28 baseball cards in the attic. eight of the cards are valuable. He decides to share the collection fairly between his four younger cousins.

? How should he divide the collection fairly?

Akeem's Solution

I will represent the eight valuable cards with yellow counters and other cards with red counters.

To be fair, I need to make smaller sets with yellow to red counters in the same ratio as in the original set. I need **equivalent ratios** for 8 : 20.

First, I will divide the cards into two sets.

To keep the comparison the same, I will split the yellow counters in half and the red counters in half.

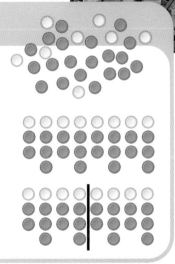

A. Write the ratio of the number of valuable cards to the number of regular cards in each of Akeem's two new sets.

B. Why is it reasonable to say that this new ratio is equivalent to 8 : 20?

C. To make four sets, divide each of Akeem's new sets into two more sets that each have a ratio equivalent to 8 : 20. Record this ratio in a ratio table.

equivalent ratios
Two or more ratios that represent the same comparison

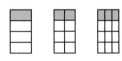

1 : 3   2 : 6   3 : 9

Number of valuable cards	8
Number of regular cards	20

D. How many valuable cards does each cousin get? How many regular cards does each cousin get? How many cards does each cousin get altogether?

Reflecting

1. How are the numbers in the ratio table related to each other?
2. Li Ming says she could have made the four equal sets by dividing each number in the original ratio by four. Do you agree or disagree? Explain.

Checking

3. Ellen has a set of stamps. The ratio of valuable stamps to regular stamps is 5 to 12.
   a) After one year her set is twice as big, but it still has the same ratio of valuable stamps to regular stamps. Compare the new number of valuable stamps to regular stamps with a ratio.
   b) Is the new ratio equivalent to the old ratio? Explain.
   c) Two years later, her stamp collection is three times as big as the original set. It still has the same ratio of valuable stamps to regular stamps. How many valuable stamps are in this set?

Practising

4. To make juice, you mix cans of concentrate with cans of water in the ratio of 2 to 6.
   a) Write an equivalent ratio of concentrate to water.
   b) Make a ratio table to represent the juice mixture.
   c) How many cans of water would you need to mix with five cans of concentrate? Explain your thinking.

5. Copy each equation. Replace the ■ with a number to make an equivalent ratio.
   a) 20 to 6 = 10 to ■
   b) 3 to 8 = 21 to ■
   c) 3 to 5 = 15 to ■
   d) 1 : 1 = ■ : 7

6. Lucia makes fruit leather to sell at a farmer's market. She uses a ratio of 6 bananas to 4 peaches.
   a) Write ratios equivalent to 6 : 4.
   b) If Lucia has 16 peaches, how many bananas does she need?
   c) If Lucia has 48 bananas, how many peaches does she need?

Bananas	3	6	9	12		
Peaches		4			10	12

CHAPTER 12

Frequently Asked Questions

Q. How do you compare and order fractions?

A. When the denominators are the same, a fraction with a greater numerator is greater because there are more identical parts of the same whole.

When the numerators are the same, a fraction with a greater denominator is less. It has the same number of parts, but the whole for the fraction with the greater denominator is broken up into more parts. Each part is smaller.

When neither the denominators nor the numerators of the fractions are the same, you might use benchmark fractions like $\frac{1}{2}$, $\frac{1}{4}$, 0 and 1 to order.

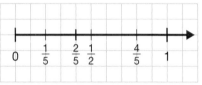

Or you might place the fractions on a number line, representing a whole by a convenient number of sections. The fraction farthest to the right is greatest.

Q. How do you write fraction and decimal equivalents?

A. Look for an equivalent fraction with a denominator of 10, 100, or 1000. Then use place value to write the decimal equivalent.

For example, $\frac{3}{5} = \frac{6}{10}$, which is read as six tenths and can be written as the decimal 0.6.

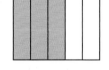

Follow place value rules to write a decimal as a fraction. Then write equivalent fractions if you want a different denominator.

For example, $0.05 = \frac{5}{100} = \frac{1}{20}$.

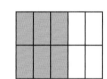

Q. How do you determine equivalent ratios?

A. Equivalent ratios represent the same comparison. You can model a ratio, then make copies of the model or break the model into equal parts.

For example, consider the ratio 6 : 4.

Model the ratio and make copies of it to get
6 : 4 = 12 : 8,

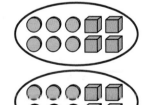

or model the ratio and divide it into equal parts to get
6 : 4 = 3 : 2

CHAPTER 12
Mid-Chapter Review

LESSON

1
1. Order each set of numbers from least to greatest. Show your work on a number line.
   a) $2\frac{1}{4}, \frac{5}{4}, \frac{3}{4}$
   b) $\frac{2}{5}, \frac{8}{5}, 1\frac{4}{5}$

2
2. Compare. Write >, <, or =. Show your work.
   a) $\frac{5}{8}$ ■ $\frac{5}{12}$
   b) $\frac{11}{4}$ ■ $\frac{5}{3}$
   c) $1\frac{1}{4}$ ■ $\frac{8}{5}$

3
3. Write each fraction as a decimal equivalent.
   a) $\frac{1}{5}$
   b) $2\frac{1}{2}$
   c) $\frac{5}{4}$

4. Write each decimal as an equivalent fraction.
   a) 0.2
   b) 0.06
   c) 0.55

4
5. In a recipe for trail mix, the ratio of nuts to dried fruits is 6 : 5.
   a) If Kylie uses 6 cups of nuts in the recipe, how much trail mix will she have altogether?
   b) Explain what ratio 5 : 11 might represent about the trail mix.
   c) Write four other ratios describing the mix of nuts and dried fruit.

5
6. Which ratios of girls to CDs are equivalent?

   A B  C D  E F

7. Luis made pie filling with blueberry and sugar in a 4 to 1 ratio.
   a) Model the ratio using counters. Sketch your model.
   b) If Luis made 5 cups of pie filling altogether, how much of each ingredient did he use?
   c) Haley adds 2 extra cups of blueberries and 1 extra cup of sugar to Luis's pie filling. What can Luis do to correct the ratio of fruit to sugar?

8. Millet is a grain. You can cook millet with 3 cups of water to 1 cup of grain. How many cups of water do you need for 7 cups of grain?

CHAPTER 12

6 Percents as Special Ratios

You will need
- a 10-by-10 grid
- a calculator

Goal Understand the meaning of percent.

James's class did a probability experiment. They threw a beach ball globe from person to person and recorded the location of the catcher's left pointer finger. This is what they recorded:

- The ratio of times the location was in the Atlantic Ocean to the total number of tosses was 4 to 25.
- The fraction of times the location was in the Pacific Ocean was $\frac{7}{20}$.

? According to the experiment, which ocean covers more of the Earth's surface?

percent
A part-to-whole ratio that compares a number or an amount to 100

$25\% = 25 : 100 = \frac{25}{100}$

James's Solution

I need to compare the ratio 4 to 25 and the fraction $\frac{7}{20}$.
If I write them both as **percents** I will be comparing to the same whole, 100.

To write the ratio 4 : 25 as a percent, I need to determine an equivalent ratio with 100 as the second number.
I can represent 4 : 25 on a hundredths grid. I'll outline 25 squares and colour 4 of them.
I can keep outlining 25 squares and colouring four until the whole grid is outlined.
16 squares are coloured. 4 : 25 is equivalent to 16 : 100.
The finger was on the Atlantic Ocean 16% of the time.

I can write 7 out of 20 as either $\frac{7}{20}$ or the ratio 7 : 20.
I can write an equivalent percent using the hundredths grid.
7 : 20 is equivalent to 35 : 100.
The finger was on the Pacific Ocean 35% of the time.

According to the experiment, the Pacific Ocean covers more of the Earth's surface than the Atlantic Ocean.

Reflecting

1. How was the grid useful in determining each percent?
2. How did writing the ratio 4 to 25 and the fraction $\frac{7}{20}$ as percents allow James to compare the results?
3. Explain how to write a ratio or fraction as a percent. Use examples.

Communication Tip

Percents are written with a percent sign (%).

The percent sign is like writing "of each 100." 25% is read "25 percent" and means "25 of each 100"

100% means the whole.

Checking

4. James researched these numbers about drylands. Canada has 3 km² of drylands for each 20 km² of total land area.
   a) Write an equivalent ratio with 100 as the whole.
   b) Write the percent of dryland in Canada.

Practising

5. Write each as a ratio, a fraction, and a percent.
   a) 12 to 100   b) $\frac{91}{100}$ c) 0.01   d) 50 out of 100

6. For each part to whole ratio, write an equivalent ratio using 100 as the whole. Use a hundredths grid. Write each ratio as a percent.
   a) 1 to 2   b) 1 : 4   c) 6 to 25   d) 18 to 20

7. What percent of each whole is coloured?
   a)   b)   c)

8. The population of teachers and students in a school is 4% teachers.
   a) Write a ratio for the number of teachers to the population.
   b) Write a ratio for the number of students to the population.
   c) Write a ratio for students to teachers.

9. Statistics Canada found that 22% of students chose baseball as their favourite sport. Faith surveyed the 25 students in her class. 12 chose baseball. Is baseball more or less popular in Faith's class than in Canada? Show your work.

10. Create and solve your own question that involves percents.

7. Relating Percents to Decimals and Fractions

You will need
- a calculator
- a 10-by-10 grid

Goal Compare and order percents, fractions, and decimals.

Rodrigo gathered newspaper clippings about the materials in a local landfill site. One article was damaged so he couldn't read part of it.

? What material takes up the most space at the landfill site?

Landfill ¼ Full of Plastic

Too Much Paper? Landfill 30% full of paper this year.

Food and yard waste pickup fills 11% of landfill site.

⅕ Landfill miscellaneous trash.

What's in Your Garbage? Rubber and leather Metal The local landfill has 2% more rubber and leather than metal.

Rodrigo's Comparison

Writing everything as a percent will make it easier to compare the numbers. I'll make a table and include a column for decimals to help me go from fractions to percents.

In the first row, 1 out of 4 is equivalent to 25 out of 100, which is 0.25 or 25%.

Material	Percent	Fraction	Decimal
Plastic		$\frac{1}{4}$	
Food and yard waste	11%		
Miscellaneous trash		$\frac{1}{5}$	
Paper	30%		
Rubber and leather	? (2% more than metal)		
Metal			
Total			

A. If everything from the landfill site is included in the table, what should be the total for the percent column? What should be the total for the decimal column?

B. Copy the table and complete the first four rows.

C. How much of the landfill is filled with rubber, leather, and metal altogether?

D. How much of the landfill is filled with rubber and leather? How much is filled with metal? Complete the table.

E. Order the materials from least amount to greatest.

F. What material takes up the most space at the landfill site?

Reflecting

1. Explain how you knew the totals for the percent column and the decimal column.
2. a) How do you write a fraction as a percent? Use $\frac{6}{8}$ as an example.
   b) How do you write a decimal as a percent? Use 0.3 as an example.
3. Did you order the amounts using the percent, fraction, or decimal column? Explain.

Checking

4. The Scouts are collecting paper for a recycling drive.
   a) Copy and complete the table.
   b) Order the types of paper from least to greatest.

Type of Paper	Percent	Fraction	Decimal
Newspaper		$\frac{1}{2}$	
Magazines	40%		
Other			0.1
Totals			

Practising

5. Write each set of numbers as percents and order them from least to greatest.
   a) 0.4, 0.04, 0.44, 0.1
   b) $\frac{1}{2}, \frac{47}{100}, \frac{3}{10}$
   c) 0.75, $\frac{1}{4}, \frac{3}{5}$, 0.85

6. Hannah researched different countries to see what part of the world's rainforest each country has.
   a) Copy and complete the table.
   b) Order the locations from least amount of rainforest to most.

Location	Fraction	Decimal	Percent
Brazil	$\frac{8}{25}$		
Indonesia		0.11	
Zaire			9%
Other			
Total			

7. The Grade 5, 6, and 7 classes are putting on a talent show. Fifty percent of the participants are from Grade 6 and 0.3 of the participants are from Grade 5. What is the fraction of participants from Grade 7?

8. About 65% of last year's campers are returning this year. Which of these statements is more accurate? Explain your choice.

> All but $\frac{3}{8}$ of last year's campers are returning this year.

> About $\frac{1}{4}$ of last year's campers are not returning this year.

CHAPTER 12

8 Estimating and Calculating Percents

You will need
- a calculator
- a 10-by-10 grid

Goal Estimate and calculate percents.

The Grade 6 class is on a 500 km bus trip to Ottawa.
It will take about 6 h.
"How much farther till we're there?" asked Isabella.
"How many more hours till we're there?" asked Khaled.
The bus driver replied, "We've gone 25% of the way."

? How many kilometres are left in the trip and how long will it take to travel that far?

Chandra's Solution

100% of the trip is 500 km. 100% of the trip will take 6 h.
I can draw a number line to help me answer the questions.

```
Start    1/4    Halfway    3/4    Ottawa
0%       25%    50%        75%    100%
|--------|------|----------|------|-->
0 km     ?      250 km     ?      500 km
0 h      ?      3 h        ?      6 h
```

50% of the trip is the same as $\frac{1}{2}$ of the trip.
One half of 500 km is 250 km. One half of 6 h is 3 h.

A. Draw a number line like Chandra's. Complete the distances for 25% of the way and 75% of the way.

B. Complete the times for 25% of the way and 75% of the way.

C. How many kilometres do they have left to go?

D. How many more hours will the bus trip take?

Reflecting

1. What strategy did you use to determine 25% of the distance and 75% of the distance?

2. Later, the class passed this sign. About what percent of the total distance have they travelled at this point?

3. Why was it helpful to know that 25% is $\frac{1}{4}$ and 50% is $\frac{1}{2}$ to solve the problem?

Checking

4. The Grade 8 students went on a 40 km canoe trip.

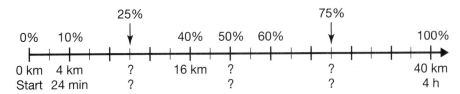

   a) The number line for the trip shows 10% of 40 km as 4 km. Explain how to determine 10% when you know 100%.
   b) Explain how to use 10% to determine 40%.
   c) Complete the number line for the trip.

Practising

5. Estimate or calculate. Show your work.
   a) 50% of the students in a class of 24 students
   b) 10% of the cost of a T-shirt if the whole cost is $12.99
   c) 25% of a 10 kg bag of sugar

6. Estimate the percent of each number. Show your work.
   a) 40% of 55   b) 5% of 160   c) 75% of 128

7. Estimate each. Show your work.
   a) 10% of a busload of 62 students
   b) 50% off shoes whose full price is $79.95
   c) 15% tip on a meal that cost $24.76

CHAPTER 12

9 Unit Rates

You will need
- calculator
- counters
- play money

Goal Represent relationships using unit rates.

Each game at the class fun fair has a different cost.

? Which game is the best bargain?

Tom's Solution

Make a Plan
I can't compare the costs of each game the way they are written, but I could compare them if I knew the cost for one play in each game.

I will calculate the **unit rate** for plays in each game.

I will use counters. Each counter will represent one play.

Carry Out the Plan
For Sponge Toss, the rate is $2.50 for 5 plays.

If I use 8 quarters for $2.00, then I show 10 quarters for 5 plays.

This is the same as 2 quarters for each play, or 50¢. The unit rate for the Sponge Toss is 50¢ for 1 play.

unit rate
A comparison of two quantities where the second one is described as 1 unit.

For example, a unit rate might be 30 km in 1 h or 4 tomatoes for $1

Communication Tip
Rates often have words like "per" or "for" in them. A slash (/) is sometimes used instead.

For example, you read 100 km/h as "100 km per hour."

A. Calculate the unit rate for one play of Super Darts and one play of Ball Toss.

B. Order the unit rates for one play from least to greatest.

C. Which game is the best bargain?

Reflecting

1. What does the word *unit* mean in the phrase *unit rate*?
2. Could Tom also have compared the prices by determining the number of plays for $1 in each game? Which way do you think is easier? Explain.
3. How is a unit rate like a ratio? How is it different?

Checking

4. Adrian wants to play the game that is the best bargain.
   a) Determine the number of plays for $1 in each game.
   b) Order the games from least unit rate to greatest unit rate.

Practising

5. Calculate the unit rate for each item.
   a) 6 batteries for $3.00
   b) 3 batteries for $1.95
   c) 2 CDs for $9.00
   d) 4 CDs for $16.00

6. Isaac found these heart rates of different animals.
   a) What is the unit heart rate in beats per minute for each animal?
   b) Order the unit rates from least to greatest.

Mammal	Heart Rate
rabbit	400 beats in 2 min
lion	40 beats in 1 min
shrew	400 beats in 30 s
elephant	140 beats in 4 min

7. a) What is the price for one of each type of cookie?
   b) Why is your answer to Part a) a unit rate?
   c) Use your unit rates to decide which cookie is the least expensive.

raisin cookies	5 cookies for $3.75
oatmeal cookies	7 cookies for $6.30
almond cookies	$4.00 for 8 cookies
carob chip cookies	$6.00 for 6 cookies

8. Who moved the fastest? Order these speeds from least to greatest.

Alexa walked 12 km in 3 h.
Alex cycled 10 km in 1 h.
Jenna cycled 8 km in 30 min.
Luke walked 10 km in 2 h.

9. A 200 g bag of chips costs $3.00 and a 300 g bag costs $4.00. Which bag is a better bargain? How do you know?

CHAPTER 12

10 Solving Problems Using Guess and Test

Goal Use a guess and test strategy to solve problems.

Cars make up 40% of Emilio's collection of 20 miniature cars and trucks. Emilio adds some cars to the collection and says, "Now I have 70% cars."

? How many cars did Emilio add to the collection?

Denise's Solution

Understand

Emilio started with 20 vehicles that made 100% of his collection.

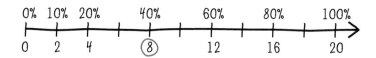

If 100% is 20 cars, then 10% is 2 cars and 40% is $4 \times 2 = 8$ cars.

20 vehicles − 8 cars = 12 trucks.

When Emilio started, the ratio of cars to trucks was 8 : 12.

Make a Plan

I will set up a table and use a guess and test strategy to determine the number of cars he added.

Carry Out the Plan

Guess	Number of cars	Number of trucks	Number of cars and trucks	Percent of cars
Original	8	12	20	$\frac{8}{20} = 40\%$
Add 10 cars	8 + 10 = 18	12	30	$\frac{18}{30} = \frac{6}{10} = 60\%$
Add 20 cars	8 + 20 = 28	12	40	$\frac{28}{40} = \frac{7}{10} = 70\%$

When I tested adding 20 cars, the percent of cars was 70%.
Emilio added 20 cars to his collection.

Reflecting

1. How did the table help Denise solve this problem?
2. How did Denise test each of her guesses?

Checking

3. Engines make up 25% of Denise's collection of 12 miniature train engines and boxcars. After Denise added new engines, her collection was 50% engines. How many engines did she add?

Practising

4. The ratio of red to blue in a purple dye is 2 : 3. How many parts of blue need to be added to change the dye to 80% blue?

5. Lily grows beans in different soil mixtures. She starts with a 3 : 4 ratio of peat moss to sand. She wants a mixture that is 60% peat moss. How many parts of peat moss does she need to add?

6. a) Identify a pair of two-digit numbers whose product contains one 9.
   b) Identify a pair of two-digit numbers whose product contains two 9s.

Math Game

Ratio Concentration

You will need
- Ratio Concentration Cards

Number of players: 2 or more

How to play: Match pairs of cards.

Assume the ratios are part to whole ratios.

Step 1   Shuffle the cards and place them face down in a 5 by 6 array.

Step 2   Take turns turning over two cards at a time.

Step 3   If the cards show equivalent ratios, keep them and take another turn. If the ratios are not equivalent, turn them face down again.

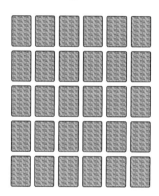

Step 4   Continue taking turns until all matches have been made.

The player with the greatest number of cards at the end wins.

Qui's Turn

I turn over these cards. The ratios are equivalent, so I keep the cards.

| 2:4 | 50% |

Next I turn over these cards. The ratios are not equivalent. My turn is over.

| $\frac{3}{5}$ | 30% |

CHAPTER 12

Skills Bank

LESSON

1

1. Name the letter on the number line that corresponds to each number.
   a) $1\frac{1}{2}$
   b) $\frac{5}{6}$
   c) $1\frac{2}{3}$
   d) $\frac{1}{6}$
   e) $\frac{5}{4}$
   f) $\frac{2}{4}$

   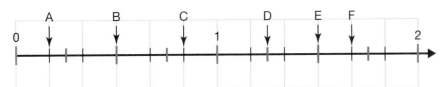

2

2. Compare. Write >, <, or =. Show your work.
   a) $\frac{4}{5}$ ■ $\frac{3}{5}$
   b) $\frac{8}{3}$ ■ $\frac{5}{3}$
   c) $\frac{1}{3}$ ■ $\frac{1}{5}$
   d) $\frac{7}{5}$ ■ $\frac{7}{4}$
   e) $1\frac{3}{4}$ ■ $1\frac{1}{2}$
   f) $2\frac{1}{2}$ ■ $2\frac{4}{8}$

3. Order each set of numbers from least to greatest.
   a) $2\frac{1}{3}, \frac{5}{3}, \frac{2}{3}$
   b) $1\frac{3}{8}, 2\frac{1}{8}, 1\frac{2}{4}$
   c) $\frac{5}{3}, 1\frac{5}{6}, \frac{2}{3}, 1\frac{1}{3}$

3

4. Write a decimal equivalent for each fraction.
   a) $\frac{2}{5}$
   b) $\frac{1}{4}$
   c) $\frac{6}{5}$
   d) $\frac{7}{2}$
   e) $1\frac{6}{8}$
   f) $2\frac{3}{4}$

5. Write an equivalent fraction for each decimal.
   a) 0.7
   b) 0.30
   c) 0.47
   d) 1.4
   e) 1.5
   f) 2.62

4

6. a) Write the ratio of glass beads to wooden beads.
   b) Write the ratio of wooden beads to glass beads.

7. Leah makes rock candy with 1 cup of water and 5 cups of sugar.
   a) Model the ratio of sugar to water with counters.
   b) Write the ratio of sugar to water.

8. Chase made Aboriginal soft-bread from a recipe that read "Mix the cups of white corn flour to sugar in a 5 to 1 ratio."
   a) If Chase used 10 cups of white corn flour, how many cups of sugar did he use?
   b) Write another part-to-part ratio for the recipe.
   c) Write four other ratios describing the flour and sugar mixture.

9. Copy and complete each equivalent ratio.
   a) 1 to 8 = ■ to 64
   b) 4 to 10 = ■ to 70
   c) 12 to 14 = ■ to 7
   d) 18 to 2 = ■ to 1

10. Sabrina wants to make a carving from plaster of Paris. She will make the plaster of Paris using a ratio of 2 parts plaster to 1 part water. If she uses 6 cups of plaster of Paris, how many cups of water will she need to add?

11. What percent of each whole is coloured?
    a)
    b)
    c)

12. Write each ratio as a percent. Use a hundredths grid.
    a) 92 to 100
    b) 12 to 20
    c) $\frac{9}{25}$
    d) $\frac{8}{40}$
    e) 8 : 100
    f) 40 : 50

13. What percent might you use to describe each?
    a) all
    b) almost all
    c) none
    d) almost none
    e) a little over half
    f) just less than one fourth

14. Write each set of numbers as percents and order them from least to greatest.
    a) 16%, $\frac{1}{10}$, 0.14, 0.9
    b) 0.98, $\frac{12}{15}$, 87%, 0.89
    c) 40%, 0.59, $\frac{6}{8}$, 0.38

15. Data was collected on how people in Toronto travel to work.
    a) Copy and complete the table.
    b) Order the methods of travel from least to greatest.

Method of travel	Fraction	Decimal	Percent
Car			72%
Walk/bicycle	$\frac{3}{50}$		
Public transit		0.22	

16. Calculate 10% and 25% of each measurement. Show your work.
    a) a 100 m swimming pool
    b) a 10 km run
    c) a 60 min class
    d) a group of 440 people

17. Estimate or calculate each percent. Show your work.
    a) 75% of 40
    b) 50% of 212
    c) 10% of $189
    d) 40% of $50
    e) 15% of 60 km
    f) 20% of 67 people

18. For each book, about how many pages are in one chapter?
    a) a three-chapter book with 45 pages
    b) a five-chapter book with 70 pages
    c) a six-chapter book with 90 pages

19. Four families are travelling to Niagara Falls from different places. Order their speeds from least to greatest.

Sapons	80 km in 2 h
Silvers	180 km in 3 h
Johns	50 km in 1 h
Cunninghams	35 km in 30 min

20. Which store has the best bargain for kiwis?

Ken's Grocery	2 kiwis for 95¢
Davis Bay Grocery	4 kiwis for $1.80
ML Variety Store	3 kiwis for $1.50

CHAPTER 12

Problem Bank

1. Jeff is $\frac{2}{3}$ of a year older than Lisa. Lisa is $\frac{1}{4}$ of a year older than Mark. If Mark is $10\frac{1}{2}$, how old is Lisa?

2. The ratio of the length of one side of a shape to its perimeter is 1 : 3. Could the shape be a triangle? a square? a rectangle? Explain your thinking.

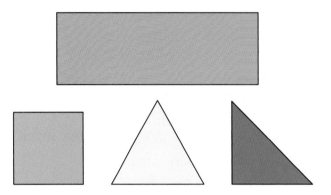

3. What are two ratios equivalent to 1 million : 1 billion?

4. A recipe for fruit punch calls for a juice to soda water ratio of 3 : 2. The ratio of fruit juices is 1 : 3 : 2 for grape to apple to raspberry. If Diana starts with 4 L of raspberry juice, how many litres of grape juice, apple juice, and soda water should she add?

5. Eight of the 20 players forming two indoor soccer teams are experienced. Once the teams are formed, team A has 60% experienced players.
   a) What is the ratio of experienced to inexperienced players on team A? Show your work.
   b) What is the percent of inexperienced players on team B? Show your work.

6. Carolyn took $64.00 spending money on a trip. At the end of the first day she had 75% of the money left. Calculate the amount of money she had spent.

7. Ivan can type 48 words each minute. About how long would it take him to type this page of the text? Explain how you calculated.

CHAPTER 12

Frequently Asked Questions

Q: What is percent?

A: Percent is a special part to whole ratio where the whole is 100.

For example, 5 students in a class say they have dogs as pets. There are 20 students in the class. The ratio of dog owners to students in the class is 5 to 20. This is a part to whole ratio.

To write this as a percent, write an equivalent ratio with 100 as the whole. 5 to 20 = 25 to 100.

Q: How do you compare percents, fractions, and decimals?

A: It is easiest to compare fractions, percents, and decimals when they are all represented as part of the same whole and written the same way: all as fractions, all as decimals, or all as percents.

If the values are fractions, you can write them as equivalent fractions with denominators 100. Then they can be compared as percents or decimal hundredths.

For example: Compare $\frac{6}{8}$, 68% and 0.7.

$$\frac{6}{8} = \frac{3}{4} = \frac{75}{100} = 75\%$$

$$0.7 = \frac{7}{10} = \frac{70}{100} = 70\%$$

From least to greatest, these numbers are 68%, 0.7 or 70%, and $\frac{6}{8}$ or 75%.

Q: What is a unit rate?

A: A unit rate is a way to compare quantities described in different ways.

For example, different sizes of a product will be priced differently. Determining unit rates for each size is a way to decide which size is the best bargain.

You can determine which size gives you the most product for $1, or which product costs the least for 1 unit.

Chapter Review

CHAPTER 12

1. As an Earth Day project, five friends started compost bins for their families. At the end of the summer, they measured how much compost they had. Order the numbers from least to greatest. Who had the most compost?

Friend	Number of bins
Jake	$\frac{2}{5}$
Kaycee	$1\frac{1}{2}$
Julie	$\frac{4}{5}$
Kim	$1\frac{1}{5}$
Gab	$\frac{7}{2}$

2. Write each fraction as a decimal equivalent.
   a) $\frac{4}{5}$ b) $\frac{3}{2}$ c) $1\frac{7}{8}$ d) $1\frac{2}{5}$

3. Write each decimal as an equivalent fraction
   a) 0.9   b) 0.75   c) 1.4   d) 2.35

4. Write the ratio of red to green in each set.
   a)
   b)

5. An orchard has 6 pear trees and 5 apple trees. Write six ratios about the trees in the orchard.

6. Paris makes trail mix with 5 cups nuts to 2 cups pretzels.
   a) Write the ratio of nuts to pretzels.
   b) Use the ratio to complete the table of values.
   c) If Paris has 18 cups of pretzels, how many cups of nuts does she need?

Nuts	10				25	15
Pretzels		8	2	12		

7. A garden of white and red roses has 60% red roses. What percent of the roses are white?

8. a) What is the ratio of Jennifer's height now to her height at age 7?
   b) What percent of Jennifer's present height is her height at age 7?

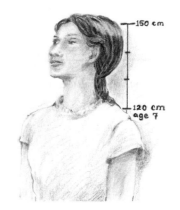

9. Alana used various sources to describe the ratio of land area on each continent to the total area of land on Earth (not including Antarctica). She used various sources.

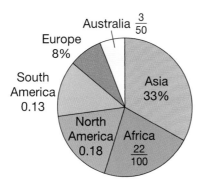

a) Copy and complete the table.

Continent	Fraction	Decimal	Percent
Asia			
Africa			

b) Order the continents from least to greatest area of land.

10. A group of 80 people went on a trip. Twenty-five percent of the travellers had 2 bags each. The rest had 1 bag. How many bags were there altogether?

11. Cole researched the average speeds of some animals. Which animal is the slowest?

Animal	Speed
snail	6 cm in 4 s
spider	38 cm in 2 s
giant tortoise	45 cm in 6 s
slug	4 cm in 8 s
centipede	100 cm in 2 s

12. A chemist made a mixture of 3 parts copper to 5 parts gold. He wants to change the mixture so that it is 85% gold. How many parts of gold must he add?

CHAPTER 12

Chapter Task

Running with Terry

Terry Fox began his run across Canada in St John's, Newfoundland in April, 1980. The map shows his route across the island.

? What fraction and percent of Terry's run was completed in Newfoundland?

Part 1: Newfoundland

A. Use string to measure the length of the line representing Terry's route through Newfoundland. Express your answer in centimetres.

B. The scale on the map shows that 1 cm on the map represents 50 km. What is the length of Terry's route in kilometres?

C. Terry ran about 35 km each day. About how many days did he take to get across Newfoundland?

Scale: 1 cm represents 50 km

Part 2: The Whole Trip

Terry ran about 5373 km in 143 days before his cancer forced him to stop in Thunder Bay, Ontario.

D. Draw a percent line from 0% to 100% showing the total time and total distance he ran.

E. Locate the times and distances for the part of the run through Newfoundland on the percent line and determine the percent of the run completed there.

F. Estimate the fraction of the total distance and the fraction of the time he spent in Newfoundland.

Task Checklist
- ✓ Did you show your calculations?
- ✓ Did you include units on your percent line?
- ✓ Did you label your percent line with benchmark percents?
- ✓ Did you check your work?

CHAPTER 13

Probability

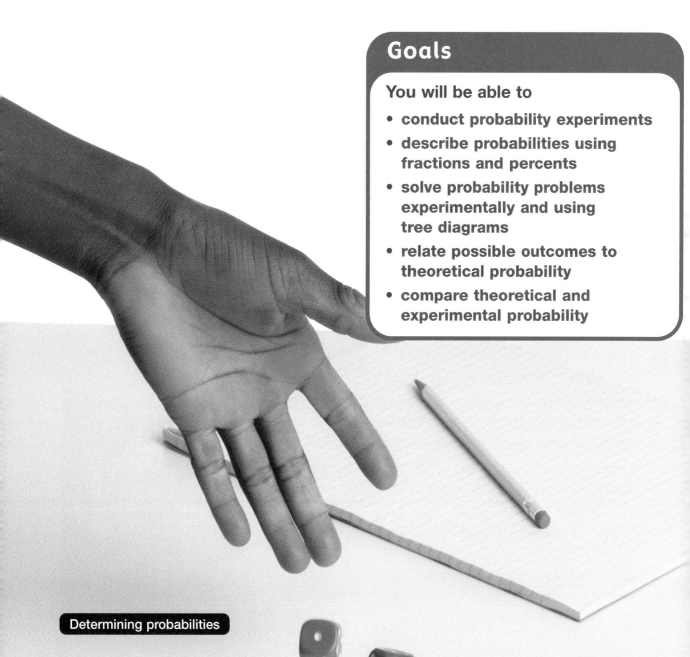

Goals

You will be able to

- conduct probability experiments
- describe probabilities using fractions and percents
- solve probability problems experimentally and using tree diagrams
- relate possible outcomes to theoretical probability
- compare theoretical and experimental probability

Determining probabilities

CHAPTER 13

Getting Started

You will need
- dice
- counters

Lucky Seven

Marc and Li Ming are playing Lucky Seven.

? **Is Lucky Seven a fair game, or is one person more likely to win than the other?**

A. Play Lucky Seven with a partner. Decide which is Player 1 and which Player 2. Make the first two rolls. Record the results of each roll.

Roll sum	Player 1	Player 2
7	✔	
10		✔

✔ shows which player won a counter

Lucky Seven Rules
1. Play with a partner.
2. Place 10 counters in a pile between you.
3. Each of you rolls a die.
4. If the sum of the dice is 5, 6, 7, or 8, Player 1 gets a counter. For any other sum, Player 2 gets a counter.
5. Repeat seven more times.

The winner is the player with the most counters at the end of the game.

B. Predict which is true:
- Player 1 is more likely to win.
- Player 2 is more likely to win.
- The two players are equally likely to win.

C. Roll another six times. Record the winner each time.

D. You rolled eight times. Use a fraction to record the probability that each player wins a counter based on your results.

E. Combine your results with the results from two other pairs. Calculate the probability that each player wins a counter using all the data.

F. How accurate was your prediction in Part B? Would you make the same prediction again? Explain why or why not.

Do You Remember?

1. Based on Tyler's data, what is the probability for each?
   a) rolling a 6
   b) rolling an even number

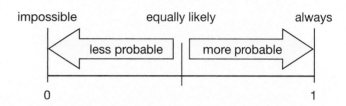

Tyler's 18 Rolls					
4	5	6	4	1	3
2	3	6	1	3	4
4	1	1	3	5	6

2. Mark each probability from Question 1 on the probability line.

3. This spinner is spun twice. Based on the results, which of the two listed events is more likely? Explain.

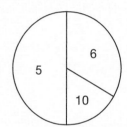

Results of Ten Double Spins					
10, 5	6, 5	5, 6	6, 5	5, 10	5, 5
10, 6	10, 5	5, 6	5, 5	10, 5	6, 6

   a) spinning two 5s **or** spinning two 10s
   b) spinning two 5s **or** spinning one 5 and one non-5

4. Julia is picking two cubes from this bag. She pulls one out, puts it back, and pulls out another one. One outcome is "blue, red". List all of the possible outcomes.

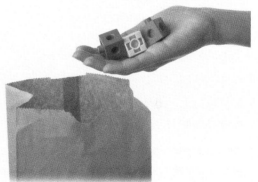

5. Sydney rolled a die 10 times. Based on her data, she listed the probability of rolling a 5 to be $\frac{3}{10}$ and the probability of rolling a 1 to be $\frac{1}{10}$. Record one possibility for what her 10 rolls might have been.

6. Record each fraction as a percent.
   a) $\frac{1}{10}$
   b) $\frac{1}{4}$
   c) $\frac{3}{10}$

CHAPTER 13

1 Conducting Probability Experiments

You will need
- dice
- counters

Goal Compare probabilities in two experiments.

You are offered a choice of two games to play.

Game 1
1. Take four counters from the "bank" of counters.
2. Pay the bank back one counter to roll.
3. Roll two dice.
4. If the sum is a multiple of 3 you get two counters from the bank.

You win if you have at least three counters after four rolls.

Game 2
1. Take 10 counters from the "bank" of counters.
2. Pay the bank back two counters to roll.
3. Roll two dice.
4. If the sum is a multiple of 4 you get five counters from the bank.

You win if you have at least eight counters after four rolls.

? Which game are you more likely to win?

A. Predict which game you are more likely to win. Justify your prediction.

B. Play Game 1 three times. Use a chart to show what happened. Based on your results, what is the probability of winning?

Game 1 Scoring

Turn	I have	I pay	I roll a sum of	I win	Now I have
1	4	1	7	0	3
2	3	1	6	2	4
3	4	1	8	0	3

C. Play Game 2 three times. Record the number of counters you have after each roll. Based on your results, what is your probability of winning?

D. How accurate was your prediction in Part A? Explain.

E. Play each game three more times.

F. Combine your data with someone else's. Which game are you more likely to win? Use probability language to explain why.

390

Reflecting

1. How many games do you think you would need to play to be fairly sure of your answer to Part F? Why?

2. In Game 1, do you think you would have ended up with more counters if you received counters for rolling a multiple of 2 rather than a multiple of 3? Explain.

3. How could you change the rules for Game 2 to make winning more likely? Explain why winning with your new rules is more likely.

Mental Imagery

Visualizing Fractions on a Number Line

You can visualize a probability on a probability line as a position between 0 and 1.

To visualize where a probability of $\frac{2}{3}$ is on the line, mentally divide the line into three equal sections and put $\frac{2}{3}$ at the end of the second section.

A probability of $\frac{2}{3}$ means that an event is likely.

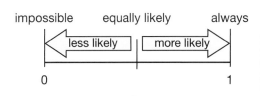

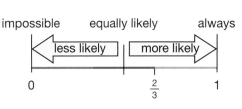

A. Why might you visualize $\frac{3}{8}$ by first visualizing $\frac{3}{4}$ and then dividing that distance in half?

B. Why might you visualize $\frac{5}{9}$ by going just a bit past $\frac{1}{2}$?

Try These

1. Place each fraction on a probability line. Tell which probability word or phrase best describes the probability.

   a) $\frac{1}{4}$ b) $\frac{4}{5}$ c) $\frac{2}{20}$ d) $\frac{5}{16}$

   - very unlikely
   - unlikely
   - likely
   - very likely

CHAPTER 13

2 Using Percents to Describe Probabilities

You will need
- spinner with 20 sections
- die

Goal Conduct experiments and use percent to describe probabilities.

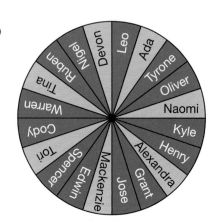

There are 20 students in a class: 13 are boys and 7 are girls. The teacher makes a spinner with all of their names on it. Each time she wants to call on a student, she spins the spinner to decide whom to call on.

? What percent describes the probability that the teacher will call on a girl?

Kurt's Strategy

I need to spin the spinner many times.

Since I want to answer with a percent, I will spin 10 times. Then the probability will be a number of tenths. Tenths are easy to rename as hundredths and then as percents.

A. Colour your spinner like the one shown. Spin it 10 times. Record your results in a tally chart.

Spin	Results
girl	|
boy	|

B. Which phrase best describes your probability of spinning a girl's name?
   - very unlikely
   - unlikely
   - very likely
   - likely

C. Place the event from Part B on a probability line.

D. Describe the probability from Part B as a fraction.

E. Describe the probability from Part B as a percent.

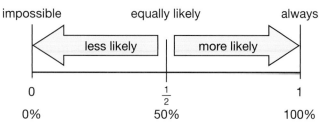

F. Conduct the experiment again. Repeat Parts B to E using the combined data for 20 spins.

G. What percent describes the probability that the teacher will call on a girl?

Reflecting

1. Why might you have expected the probability for calling on a girl to be less than 50% but greater than 25%?

2. What percents might describe each of these probabilities?
   a) impossible   b) equally likely   c) very likely

3. Why might someone choose to use percents instead of fractions to describe probabilities?

Checking

4. Describe the probability that the teacher will call on a boy using words, a fraction, and a percent. Use your data from the experiment.

Practising

5. Roll a die 20 times. Record each roll and the probability of each event as a percent.
   a) rolling 5
   b) rolling an even number
   c) not rolling a 4, 5, or 6
   d) rolling a 10

6. Robert reported that the probability of winning a certain game was 30%. How many times would you expect to win in each of these situations?
   a) if you play 100 times
   b) if you play 50 times
   c) if you play 10 times
   d) if you play 25 times

7. Predict whether each probability is closer to 10%, 50%, or 90%. Then test by rolling a die 20 times and reporting the probability as a percent.
   a) the probability of rolling a number less than 6
   b) the probability of rolling a 2
   c) the probability of rolling a number greater than 4

8. Jaspreet and Kaden are playing a game. What should Jaspreet's probability of winning be if the game is fair? Explain.

9. Make up an experiment with dice where you predict the probability of a particular event as about 30%. Test your prediction with an experiment.

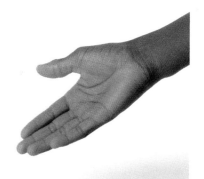

CHAPTER 13

3 Solving a Problem by Conducting an Experiment

You will need
- a play coin
- a counter

Goal Use an experiment as a problem solving strategy.

Tara made up an activity for her birthday party. There are prizes at the end of some paths of a maze, but not other paths. There are four prizes and eight paths.

? What is the probability that a party guest will end up with a prize?

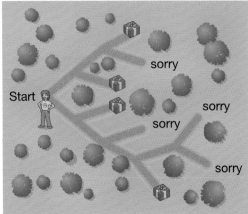

Tara's Method

Understand the Problem
I need to calculate the fraction of the time I could expect to win a prize. I predict that I'll win half the time or 50%, since I know there are four prizes and eight paths.

Make a Plan
I'll conduct an experiment. I'll flip a coin each time I come to a place in the path where I have to choose whether to go left or right. If it's heads, I'll go left. If it's tails, I'll go right. This way, the choice is **random**.

For example, if I get heads, heads, tails, I follow the red path and I don't get a prize.

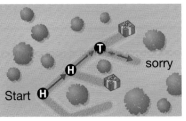

I'll do the experiment 20 times and see what fraction of the time I get a prize.

random
A result is random if what happens is based on chance. Something that is not random has to happen a certain way.

For example, the day after Tuesday is always Wednesday. That's not random.

If you put the names of the days of the week in a bag and pick Tuesday out, it is random which day you will pick next.

394 NEL

Carry Out the Plan

These are my results. Sometimes I knew if I would get a prize after only two tosses. Sometimes it took three, or four, or five tosses before I was sure.

1st toss	2nd toss	3rd toss	4th toss	5th toss	Win a prize?
H	H	T			
T	H				✔
H	H	T			
H	T				✔
T	H				✔
H	H	H			✔
T	T	H			
H	T				✔
T	H				✔
T	T	H			

1st toss	2nd toss	3rd toss	4th toss	5th toss	Win a prize?
H	T				✔
H	H	H			✔
T	T	T	H	H	
H	T				✔
T	H				✔
T	T	H			
T	T	H			
H	T				✔
H	H	T			
T	T	T	T		✔

H means turn left, T means turn right, ✔ means I win a prize.

The probability that I get a prize is $\frac{12}{20}$. That's equivalent to $\frac{60}{100}$, so the percent is 60%.

Look Back

I won more than half the time. My prediction was a bit low.

Reflecting

1. Why was a coin flip a good model to make the path choice random?
2. Is an experiment a good way to solve this problem? Explain.

Checking

3. Predict the probability of winning a prize if the faded paths are removed.
   Try the experiment using 20 trials again.
   Compare your results with your prediction.

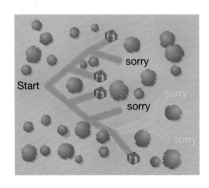

Practising

4.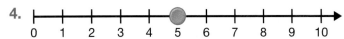

   Start at 5 on a number line. Conduct an experiment by tossing a coin 4 times. Go left if you toss heads; go right if you toss tails. Repeat the experiment a total of 20 times. What is the probability that you end up on 5?

5. Start with a counter in the centre square.
   Conduct an experiment to determine the probability that you end up in the yellow section of the grid after one turn. The counter moves four times in a turn.
   Toss the first coin. Move right if it's heads or left if it's tails.
   Toss the second coin. Move up if it's heads or down if it's tails.
   Repeat the two tosses.
   Record whether or not you end up in the yellow section.

   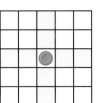

6. Kurtis's mom tosses a coin to decide who has to do the dishes each night, Kurtis or his sister. Heads means Kurtis does the dishes and tails means his sister does.
   Conduct an experiment to determine the probability that Kurtis does the dishes more than 3 times in the next week.

Curious Math

Random Numbers and Letters

When you ask a person to randomly pick a number from 1 to 10, you would expect that each number would have a probability of $\frac{1}{10}$ of being selected. But is that what actually happens?

1. Complete an experiment by asking 20 people to randomly pick a number between 1 and 10.

2. What probabilities would you suggest are more reasonable for the probability of selecting each number?

3. Suppose you ask people to choose their favourite letter of the alphabet. What fraction would describe the probability of any letter if each choice were equally likely?

4. Predict what probabilities you might get if you actually conducted this letter experiment.

5. Conduct the experiment.

CHAPTER 13

Frequently Asked Questions

Q: How can you use an experiment to decide which of two events is more likely?

A: Perform the experiment a number of times and observe which event happens more often.

For example, is it more likely to spin two 3s on this spinner, or a 5 and a 3?

You and a friend might each spin 15 times. You both get 3s $\frac{2}{15}$ of the time. One of you spins 3 and the other spins 5 $\frac{3}{15}$ of the time. The second event seems more likely, but you should probably try more spins since the probabilities are close.

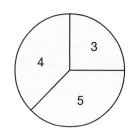

Trial spins

4, 5	3, 5	3, 3	5, 4
5, 3	5, 5	3, 4	3, 5
4, 5	5, 4	4, 3	4, 5
4, 5	3, 3	4, 4	

Q: How do you describe the probability of an event as a percent?

A: Use a fraction to describe the probability of the event.

$$\text{Probability} = \frac{\text{Number of times the event happened}}{\text{Number of times you tried it}}$$

Then calculate an equivalent fraction with a denominator of 100 and rename this as the equivalent percent.

For example, suppose you want to know the probability of rolling a total of 4 or 5 on two dice. You conduct an experiment 20 times and the sum is either 4 or 5 for 6 of those 20 tries. The probability of 4 or 5 is $\frac{6}{20}$. Rewrite $\frac{6}{20}$ as the equivalent fraction $\frac{30}{100}$. The probability of rolling 4 or 5 is 30%.

It is useful to use a percent so that you can compare probabilities for events that you tried different numbers of times. For example, if one person's probability was $\frac{6}{20}$ and another's was $\frac{10}{25}$, it's hard to compare, but it's easy to compare 30% $\left(\frac{6}{20}\right)$ to 40% $\left(\frac{10}{25}\right)$.

CHAPTER 13
Mid-Chapter Review

LESSON 1

1. Read the rules for Game 1 and Game 2.

   Game 1
   1. Take four counters from a bank of counters.
   2. Return one counter to the bank to roll the dice.
   3. Roll two dice.
   4. If the sum is 8, 9, or 10, you get two counters from the bank.

   You win if you have at least four counters after three rolls.

   Game 2
   1. Take four counters from a bank of counters.
   2. Return one counter to the bank to roll the dice.
   3. Roll two dice.
   4. If the sum is 3, 4, or 5, you get three counters from the bank.

   You win if you have at least four counters after four rolls.

   a) Predict which game has a greater probability for winning. Explain your prediction.
   b) Test your prediction by playing each game 10 times.

LESSON 2

2. Describe each probability as a percent.
   a) $\frac{4}{10}$
   b) $\frac{12}{20}$
   c) $\frac{16}{25}$

3. Describe a number or numbers you might spin on the spinner with each probability.
   a) about 10%
   b) about 25%
   c) about 50%

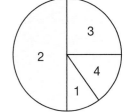

4. Roll two dice 20 times. Record your results in a chart like this.

1st die	2nd die	Is the sum 4?	Are both numbers odd?	Are the numbers 3 apart?	Is one even and one odd?

   Write each of these probabilities as a percent.
   a) rolling a sum of 4
   b) rolling two odd numbers
   c) rolling numbers three apart
   d) rolling one even and one odd

LESSON 3

5. Conduct an experiment to predict the probability of ending up on the 5th floor after 3 elevator rides, starting on the 4th floor. On each elevator ride, you go up or down only one floor. You use a coin to randomly decide whether to take the elevator up or down.

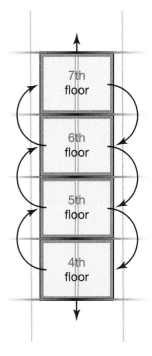

CHAPTER 13

4 Theoretical Probability

Goal: Create a list of all possible outcomes to determine a probability.

Angele and Akeem each roll a die. Angele wins if she rolls a 4. Akeem wins if the sum of both of their rolls is 4. Otherwise, it's a tie.

? Why is Angele more likely to win than Akeem?

Angele's Method

Instead of doing an experiment, I'll make a chart to show all of the possible **outcomes**. Each outcome is just as likely as any other, so I'll make the same size box for each. I'll write W if I win, L if I lose, and T for a tie.

		\multicolumn{6}{c}{Angele's roll}						
		1	2	3	4	5	6
Akeem's roll	1	T	T	L	W	T	T
	2				W		
	3				W		
	4				W		
	5				W		
	6				W		

> I roll 4 and Akeem rolls 2.

There are 6 rows, one for each possible roll for Akeem, and 6 columns, one for each possible roll for me.
There are 36 possible outcomes in the chart.
The only L in the first row is if I roll 3 and Akeem rolls 1.
The only Ws are in the column under my roll of 4.
These are the favourable outcomes for the event 'I win.'

The **theoretical probability** that I will win is $\frac{6}{36}$.

theoretical probability
The probability you would expect when you analyze all of the different possible outcomes.

For example, the theoretical probability of flipping a head on a coin is $\frac{1}{2}$, since there are 2 equally likely outcomes and only 1 is favourable.

Experimental probability is the probability that actually happens when you do the experiment.

A. Complete the chart. Use W if Angele wins, L if she loses, and T if she ties.

B. Is Angele more likely to win than Akeem? Explain.

Reflecting

1. Why did Angele say that the theoretical probability of her winning on any roll is $\frac{6}{36}$?
2. How could you have predicted that there would be 36 outcomes?
3. How do you know that the chart includes all possible outcomes for the game?

Checking

4. If you roll two dice, what is the theoretical probability of each of these events?
   a) sum of 8
   b) sum of 7
   c) sum of 3

		First Roll					
		1	2	3	4	5	6
Second roll	1	2	3				
	2	3	4				
	3	4					
	4	5					
	5	6					
	6	7					

Practising

5. If you roll two dice, what is the theoretical probability of each event?
   a) difference of 3   b) difference of 1   c) difference of 0

6. Imagine spinning this spinner twice.
   a) Does the chart show all of the possible outcomes? Explain.
   b) How could you have predicted there would be 9 outcomes with two spins?
   c) What is the theoretical probability that the sum of the two spins is an even number?
   d) What is the theoretical probability that the sum is greater than 2?

		First spin		
		1	2	3
Second spin	1			
	2			
	3			

7. A computer randomly chooses a 2-digit counting number between 1 and 100. What is the theoretical probability of each event?
   a) The number is less than 50.
   b) The number is even.
   c) The number is a multiple of 5.
   d) The ones digit of the number is greater than the tens digit.

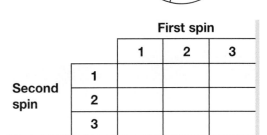

CHAPTER 13

5 Tree Diagrams

You will need
- counters

Goal Use a tree diagram to determine a theoretical probability.

Jorge and Isabella are playing a board game with a die. Jorge is two spaces behind Isabella. They each roll a die and move forward that many spaces.

? What is the theoretical probability that Jorge and Isabella land on the same space after one roll each?

Jorge's Method

I'll draw a **tree diagram** to show all the possible outcomes when we each roll once. I'll move counters to see what space we each land on.

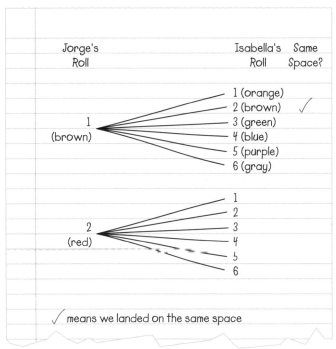

tree diagram
A way to record and count all combinations of events, using lines to form branches. This tree diagram shows all the things that can happen if you flip a coin twice.

1st flip   2nd flip

H — H (HH)
   \ T (HT)

T — H (TH)
   \ T (TT)

402

NEL

A. Copy and complete Jorge's tree diagram to show all of the outcomes. How could you predict that there would be 36 branches?

B. For how many of those outcomes did Jorge and Isabella land on the same space?

C. What is the theoretical probability that the two will land on the same space?

Reflecting

1. Why does each branch of the tree represent one of the possible outcomes?
2. The probability of landing on the same space can be expressed as a fraction. How do you use the tree diagram to determine the denominator? The numerator?

Checking

3. Determine the theoretical probability that Jorge and Isabella will land on spaces next to each other. Use the tree diagram.

Practising

4. a) Use a tree diagram to list the possible outcomes if the spinner is spun twice.
   b) Determine the theoretical probability that the difference of the numbers is 1.
   c) Determine the theoretical probability that the product of the numbers is 4.

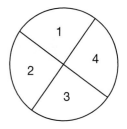

5. Randomly choose two different vertices of the hexagon.
   a) Determine the theoretical probability that a line segment joining them is longer than the side length of the hexagon. Use a tree diagram.
   b) Determine the theoretical probability that the letters of both vertices come before E in the alphabet.

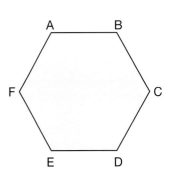

6. Use a tree diagram to show why the Lucky Seven game in Getting Started on page 388 is not fair.

403

CHAPTER 13

6 Comparing Theoretical and Experimental Probability

You will need
- coloured cubes
- paper bag
- dice

Goal Compare the theoretical probability of an event with the results of an experiment.

One yellow cube, one blue cube, and two red cubes are placed in a bag. They are mixed up and two cubes are pulled out, one at a time, without looking.

? What is the probability of pulling out a red cube then a blue one?

Maggie's Explanation

I can determine the theoretical probability if I draw a tree diagram or I can do an experiment to get the experimental probability.

I'll do the tree diagram first.

I'll label the first red cube R1 and the second one R2.

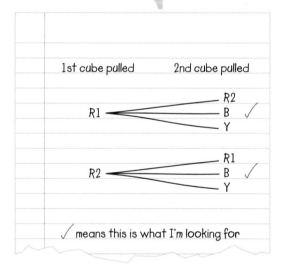

1st cube pulled   2nd cube pulled

R1 — R2
    — B ✓
    — Y

R2 — R1
    — B ✓
    — Y

✓ means this is what I'm looking for

A. Copy and complete the tree diagram to show all possible orders for drawing the cubes.

B. What is the theoretical probability of drawing a red cube and then a blue one?

C. Perform an experiment where you pull two cubes from a bag like the one described above. Do not put back the first cube before you pull the second one. Repeat the experiment 10 times. Determine the experimental probability that a red cube is chosen and then a blue cube.

	First pull	Second pull
1	Y	R
2	R	B
3		

D. Combine your data with the data from other groups. Determine the new experimental probability that a red is chosen before a blue.

Reflecting

1. When you performed the experiment in Part C, why could you stop if the first cube isn't red?

2. How did the experimental and theoretical probabilities compare when you used your own data? What about when you used the combined data?

Checking

3. There are two red cubes, a blue cube, and a yellow cube in a bag. Compare the theoretical probability and the experimental probability of choosing two different-coloured cubes.

Practising

4. a) What is the theoretical probability of rolling an even number on a die?
   b) Roll a die 20 times. Compare the experimental probability of rolling an even number to the theoretical probability.

5. a) What is the theoretical probability that, when you roll two dice, both numbers will be a multiple of 3?
   b) Conduct an experiment with at least 20 rolls. What is your experimental probability for this event?
   c) Why might the experimental probability be different from the theoretical probability?

6. a) What is the theoretical probability that, when you roll a die twice, you will roll a number greater than 3 before you roll a number less than 3?
   b) Compare this to an experimental probability. Repeat your experiment at least 20 times.

Math Game

No Tails Please!

You will need
- play coins

Number of players: 2

How to play: Flip a coin. Get a point for each head. Stop when you want.

Step 1   Flip a coin. You win 1 point for each head. You can flip as often as you wish.

Step 2   Your turn ends when you flip a tail, or you choose to stop for the turn.

Step 3   If you flip a tail, you lose all your points for the turn.

Step 4   If you stop before you flip a tail, add the points from the turn to the points from earlier turns.

The game ends when a player has 10 points.

James's Turn

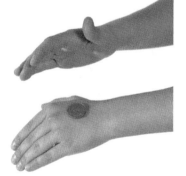

I have 3 points already.

This turn I flip a head, then another head.

Now I have 5 points. I think I'll stop before I get a tail!

Turn	Flips	Turn points	Total points
1	T	0	0
2	H H H	3	3
3	H T	0	3
4	H H	2	5
5			

CHAPTER 13

Skills Bank

LESSON

1

1. Use a fraction to describe each probability for Keith's rolls of a die.

   Keith's Rolls of a Die

   First 10 rolls
   4 5 2 3 1 2 6 4 2 3

   Next 10 rolls
   2 1 1 3 1 6 5 4 6 4

   Next 10 rolls
   1 2 3 6 4 5 1 2 1 3

   a) probability of an even number in the first 10 rolls
   b) probability of an odd number in all 30 rolls
   c) probability of a number less than 3 in the first 20 rolls

2. The winners of two different games are shown in the tables.

   Game 1

Player A	✓				✓	✓	✓		✓	
Player B		✓	✓	✓				✓	✓		✓

   Game 2

Player A			✓		✓		✓				
Player B	✓	✓		✓		✓		✓	✓	✓	✓

   a) What is Player A's probability of winning each game?
   b) What is Player B's probability of winning each game?
   c) Which game seems to be more fair?

2

3. Use percents to describe the probabilities for Bridget's spins.
   a) spinning a 2
   b) spinning a 5
   c) spinning a 1
   d) spinning an even number

   Bridget's 25 Spins

2	3	3	1	4
1	2	5	1	2
2	3	1	1	2
4	5	2	4	1
1	2	3	3	1

4. a) Repeat Question 3 using only the first 10 spins.
   b) How did the probabilities change?

5. Roll two dice 20 times. Record the result of each roll. Use percent to describe the probability for each.
   a) two numbers greater than 3
   b) two prime numbers
   c) a sum of 3

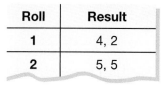

Roll	Result
1	4, 2
2	5, 5

6. Holden put a counter at 6 on a number line. He flipped a coin to decide how to move. He moved left for heads and right for tails. Conduct an experiment. What is the probability that Holden will end on 7 after three moves?

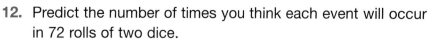

7. Use the same number line as in Question 6. What is the probability that Holden will end on 4 after four moves? Conduct an experiment.

8. What is the theoretical probability of each when rolling a die?
   a) a factor of 6   b) a multiple of 2   c) a multiple of 1

9. What is the theoretical probability of each when rolling two dice?
   a) 4 and 3
   b) 2 and another even number
   c) two consecutive numbers
   d) a product of 3
   e) a product of 6
   f) numbers 3 apart

10. a) Draw a tree diagram to show all possible outcomes when spinning this spinner twice.
    b) Determine the probability of spinning the same number twice. Use the tree diagram.
    c) Determine the probability of spinning two numbers with a difference less than 3. Use the tree diagram.

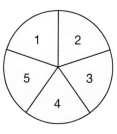

11. Draw a tree diagram to determine the probabilities that when you pick two cubes from the bag, you pick the red cube before a blue one. Once you pick a cube, you don't return it.

12. Predict the number of times you think each event will occur in 72 rolls of two dice.
    a) sum of 6   b) difference of 2   c) product of 8

CHAPTER 13

Problem Bank

LESSON

2

1. Use a percent to describe the probability that you will spin two 3s if you spin this spinner twice. Test your prediction with an experiment.

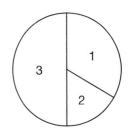

2. Roll a die twice to give you the coordinates of a point on the grid. Repeat twice more and form a triangle (or a line) using all three pairs. Create at least 20 triangles (or lines). What is the probability, as a percent, that the coordinates form an isosceles triangle?

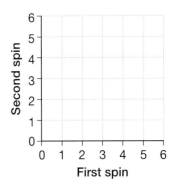

3. Describe two situations for which the probability of the second situation is about 20% greater than the probability of the first one.

3

4. Imagine that a spider lands on one square randomly. The spider then moves randomly through a side of the square into another square. Conduct an experiment to determine the probability that the spider will end up on the pink square.

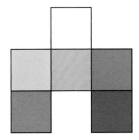

5. Create a prize maze like the one on page 394, where you predict that the probability of winning a prize is about 25%. The maze needs at least six paths. Test your prediction with an experiment.

6. Conduct an experiment to determine the probability that you will pull out the blue cube before a pink one if you pull out three cubes. Don't replace a cube once you've pulled it out.

7. What is the theoretical probability that, if you spin twice, the sum will be greater than 4?

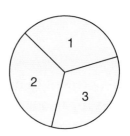

8. Three dice have these nets.
   If all three dice are rolled, which is most likely to show the greatest number? Explain.

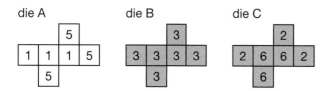

9. You roll three normal dice and score one point for each 3 or 6 that shows.
   a) What are the possible scores?
   b) Are all of the possible scores equally likely? Explain.

10. If you select two different vertices of the cube at random, what is the probability that a line segment joining them will be inside the cube rather than on a face or an edge?

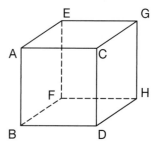

11. A tree diagram has 18 branches. It describes the possible outcomes when you roll a die and spin a spinner. What might the spinner look like? Explain why.

12. Todd rolls two dice.
    He gets 3 points if his total roll is 3 or 4.
    He gets 2 points if his total roll is 7.
    a) Is Todd more likely to get 3 points or 2 points? Explain.
    b) Conduct an experiment and compare the experimental results to the theoretical probability.

CHAPTER 13

Frequently Asked Questions

Q: What is theoretical probability and how is it different from experimental probability?

A: Theoretical probability is the probability that you would expect when you consider all of the possible outcomes.

$$\frac{\text{Number of favourable outcomes for the event}}{\text{Number of possible outcomes}}$$

For example, the theoretical probability of rolling a 2 on a die is:

$$\frac{1 \text{ favourable outcome for the event 2}}{6 \text{ equally likely outcomes}} = \frac{1}{6}$$

Q: How can you use a tree diagram to determine theoretical probability?

A: A tree diagram is a tool used to make it easier to count all of the possible outcomes when two or more things are happening.

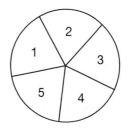 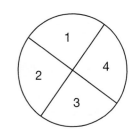

Spinner 1   Spinner 2

For example, the tree diagram shows that there are 20 possible outcomes when these two spinners are spun.

The probability that the difference between the spins is 1 is shown by comparing the number of favourable outcomes (outcomes with checkmarks) to the total number of possible outcomes.

$$\frac{7 \text{ favourable outcomes}}{20 \text{ equally likely outcomes}} = \frac{7}{20}, \text{ or } 35\%$$

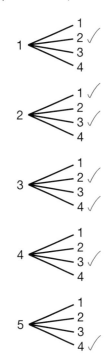

CHAPTER 13
Chapter Review

1. Which game do you predict gives Player A the greater probability of winning? Test your prediction by playing.

   Game 1
   1. Roll a die three times and add the values rolled.
   2. Player A gets a point if the sum is 7 or less.
   3. Player B gets a point if the sum is greater than 7. The first player to 10 points wins.

   Game 2
   1. Roll a die four times and add the values rolled.
   2. Player A gets a point if the sum is 12 or less.
   3. Player B gets a point if the sum is greater than 12. The first player to 10 points wins.

2. Put three red and two blue cubes in a bag.
   a) Use a percent to predict the probability that if you pull two cubes, both will be red.
   b) Conduct an experiment to test your prediction.

3. Which percent best estimates each probability, 20%, 50%, or 100%? Explain your choice.
   a) probability that a whole number less than 100 is a multiple of either 3 or 7
   b) probability that a whole number less than 100 is a multiple of 1
   c) probability that a whole number less than 100 has a 3 as one of its digits

4. Wendy picked a single cube out of a bag, and recorded which colour she picked. What percent describes Wendy's probability of choosing a black cube?

Wendy's cube picks

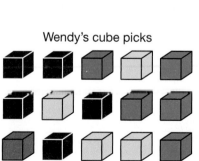

5. Conduct an experiment to determine the probability that the first word of the first numbered question on a page of a math book begins with a vowel.

6. The names of the days of the week are written on seven separate slips of paper in a bag. You pull one slip out. Describe the theoretical probability for each.
   a) The day is a weekday.
   b) The day begins with the letter T.
   c) The day is within 3 days of a Saturday.

7. What is the theoretical probability of each?
   a) rolling at least one 6 when rolling two dice
   b) rolling at least one even number when rolling two dice

8. Describe a situation where the theoretical probability is $\frac{4}{6}$.

9. Draw a tree diagram to determine the probability that, when you spin twice, one number is not a factor of the other.

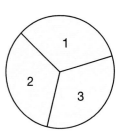

10. Why is a tree diagram a useful tool for determining a theoretical probability?

11. a) Determine the theoretical probability that, if you roll a die twice, the numbers will be in increasing order.
    b) Predict the number of times this will happen in 36 double rolls of a die.
    c) Conduct an experiment. Compare the number of times your rolls were in increasing order to your prediction. Explain why a theoretical prediction might be different from what actually happens.

CHAPTER 13

Chapter Task

Winning Races

Jane, Alyson, and Keeley are the best runners in their school. They always come in first, second, or third in races. Each comes in first in a 2 km run about $\frac{1}{3}$ of the time.

? What is the probability that Alyson will win both 2 km runs in the June competitions?

A. Draw a tree diagram to determine Alyson's probability of winning at least one of the runs.

B. Use your tree diagram. What is Alyson's probability of winning both runs?

C. Explain why the experiment described is a fair way to determine the experimental probability that Alyson wins both runs.

D. Conduct the experiment in Part C.

E. How do the probabilities in Parts B and D compare? Which would you use to describe the probability? Why?

F. Suppose Alyson trains really hard so she can win $\frac{1}{2}$ of her races. The other two girls would each win only $\frac{1}{4}$ of the time. Would the tree diagram you drew for Parts A and B help you determine Alyson's new probability of winning both races? Explain.

G. Conduct an experiment to calculate the percent probability that Alyson will win both races after training hard.

My Running Experiment

Roll a die. If the result is 1 or 2, Jane wins.
If the result is 3 or 4, Alyson wins.
If the result if 5 or 6, Keeley wins.
Roll twice. See if Alyson wins both times.
Repeat the experiment 25 times.
Calculate the percent probability that Alyson wins both runs.

Task Checklist

☑ Did you make sure your tree diagram represented all possible outcomes?

☑ Did you conduct the experiments an appropriate number of times?

☑ Did you explain your thinking?

☑ Did you support your conclusions?

CHAPTER 14

Patterns and Motion in Geometry

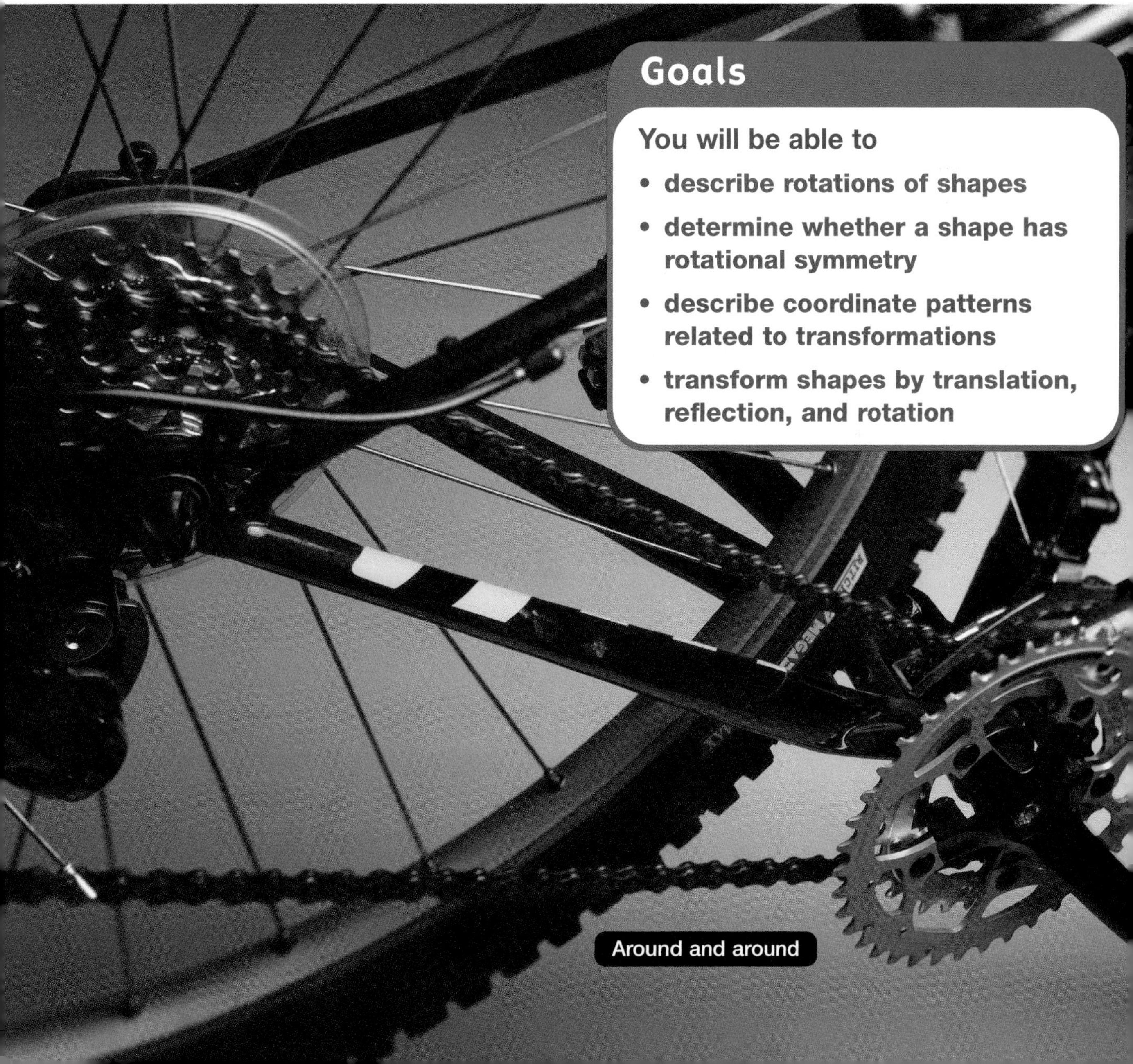

Goals

You will be able to

- describe rotations of shapes
- determine whether a shape has rotational symmetry
- describe coordinate patterns related to transformations
- transform shapes by translation, reflection, and rotation

Around and around

CHAPTER 14

Getting Started

Creating a Design Using Transformations

You will need
- pattern blocks
- grid paper
- a ruler
- a protractor
- a transparent mirror

Maggie is creating a design for the border of a greeting card. She uses translation and reflection of a shape.

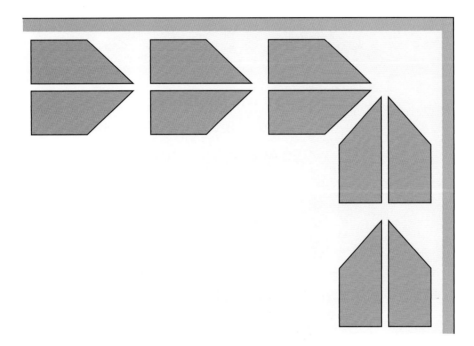

? **How can you create a design using translation and reflection?**

A. Trace Maggie's design.

B. Describe how Maggie used translation in her design.

C. Describe how Maggie used reflection in her design. Use a transparent mirror to identify the lines of reflection.

D. Create your own basic shape using pattern blocks.

E. Create a design by translating and reflecting the basic shape.

F. Describe the transformations you used.

Do You Remember?

1. Draw a 90° angle.

2. Draw each shape.
   a) an equilateral triangle
   b) a trapezoid
   c) a kite
   d) a parallelogram

3. a) Describe a translation to move the blue star onto the green star.
   b) Describe a translation to move the blue star onto the yellow star.

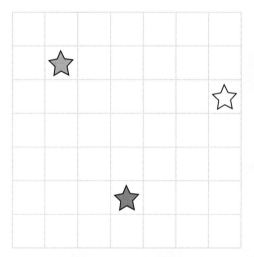

4. Draw the stars from Question 3 on grid paper.
   a) Reflect the blue star so it is in the same column as the yellow star. Draw the line of reflection. Use a transparent mirror.
   b) Reflect the blue star so it is in the same column as the green star. Draw the line of reflection.
   c) Reflect the blue star so it is in the same row as the yellow star. Draw the line of reflection.

CHAPTER 14

1 Describing Rotations

You will need
- a protractor
- Art

Goal Perform and describe the rotation of a shape around a centre that is on the shape.

Raven is looking at a picture in a book.

Raven's Pictures

I see the tail of a fish swimming to the right.

I'll use **rotations** of the book to see other pictures.

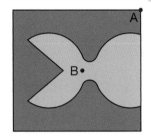

Point A is my **centre of rotation**. I can rotate the picture in a **clockwise (CW)** direction or a **counterclockwise (CCW)** direction about point A. Then I can measure the angle of rotation.

To measure the angle of rotation, I put my protractor on the centre of rotation and measure how the top side moved.

This is a 90° rotation CW about point A. It looks like the tail of a mermaid.

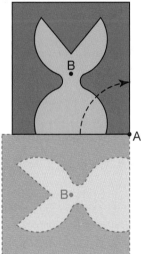

rotation
A turn of a shape described using the angle of rotation, the direction, and the centre of rotation

centre of rotation
The point that a shape rotates around

clockwise (CW)
The direction the hands of an analog clock move

counterclockwise (CCW)
The opposite direction from clockwise

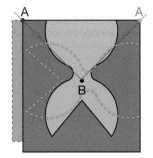

? **How can rotations change the picture that you see?**

A. Raven rotated the picture from its original position about point B. What is the direction of the rotation? What is the angle of the rotation?

B. Rotate the picture around points A and B. Use 90° and 180° as angles of rotation. Try CW and CCW directions.

C. Describe each of the pictures you see. How many different pictures are there?

Reflecting

1. Compare the rotations about the two centres A and B.
   a) How are they the same?
   b) How are they different?

2. a) Which angles of rotation give the same results when rotating in CW and CCW directions?
   b) Which angles of rotation give different results when rotating in CW and CCW directions?

3. Raven saw the tail of the fish. Imagine that she kept turning the picture 90° CW around point B over and over. What pattern of pictures would she see?

Checking

4. Describe rotations about points A and B that will turn this arrow upside-down.

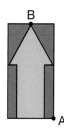

Practising

5. Describe a rotation that leaves this picture looking the same.

6. a) How can you rotate this E to get a 3?
   b) How can you rotate the picture to show another letter? Describe the rotation.
   c) Choose a different direction and centre of rotation that will show the same second letter. Describe the rotation.
   d) How can you rotate the design to show a third letter? Describe the rotation.

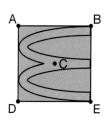

7. a) Create a picture that looks the same when you rotate it.
   b) Create a picture that looks different when you rotate it. Describe a rotation that makes the picture look different.

CHAPTER 14

2 Performing and Measuring Rotations

You will need
- a protractor

Goal Perform and describe rotations of shapes about centres not on the shape.

It's raining and Jorge notices that a leaf is caught in the windshield wiper. The leaf is moving with the wiper.

? How can the movement of the leaf be described as a rotation?

Jorge's Description

The windshield wiper is rotating about a point. I'll label it as point A.

Since the leaf moves with the wiper, I will use point A as the centre of rotation for the leaf's movement.

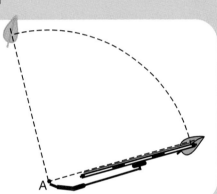

A. Measure the angle of rotation about point A.

B. Describe the rotation by naming the angle, direction, and centre of rotation.

C. Rotate the leaf 90° CCW about point A from its current position. Draw a diagram to show this rotation.

Reflecting

1. Did the distance between the leaf and point A change during its movement on the windshield? Explain.

2. How do you rotate a shape about a centre that is not on the shape?

Checking

3. Marc rotated the faded shape to its new position.
   a) Which point is the centre of rotation?
   b) Describe the rotation by naming the angle, direction, and centre of rotation.
   c) Draw the new location of the shape if it is rotated again around the same centre in the same direction with the same angle of rotation.

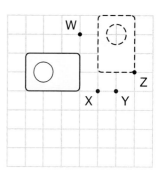

Practising

4. Debra waves a flag at a parade back and forth.
   a) Describe the flagpole's movement in both directions as a rotation.
   b) Debra waves her flag with a smaller angle of rotation. Draw and describe a possible rotation Debra might make.

5. A pendulum on a grandfather clock swings back and forth.
   a) Describe its movement as a rotation.
   b) Draw a diagram showing the movement when the angle of rotation is decreased.

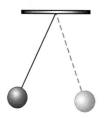

6. A pony walked along part of a circular track.
   a) Describe the pony's movement as a rotation.
   b) Describe another route it could have taken to get to the same spot.

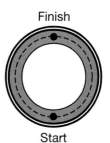

421

CHAPTER 14

3 Rotational Symmetry

You will need
- pattern blocks
- scissors
- tracing paper

Goal Determine whether and how a shape can be turned to fit on itself.

Li Ming is putting the cover back on a box.

? How many ways can Li Ming fit the cover back on the box?

Li Ming's lid

I'll try rotating a block that is the same shape as the lid.

I trace the square block. Then I put a sticker at the top left corner of the block to keep track of the rotations. I draw a black dot for the centre of rotation.

I predict that when I rotate the block around its centre, I will be able to fit the block inside the tracing four different ways, and then the sticker will return to the top left.

I test my prediction by rotating the block until it fits the tracing.

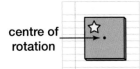

centre of rotation

I rotate the block again until it fits the tracing again.

422

NEL

I repeat the rotations until the sticker returns to the original position.

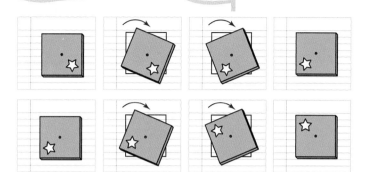

I know the square has done one complete rotation around its centre because the sticker is back to the top left corner.

The square can fit on itself four times during one complete rotation. My prediction was correct.
The square has **order of rotational symmetry** of 4.

I can fit the lid on the top of the box four ways with a different side at the front each time.

rotational symmetry
A shape that can fit on itself exactly more than once in one complete rotation has rotational symmetry.

order of rotational symmetry
The number of times a shape will fit on itself exactly during one complete rotation

Reflecting

1. How did putting a sticker at one of the corners of the block help keep track of the rotation?

2. Which of these lids has rotational symmetry? What is the order of that symmetry?

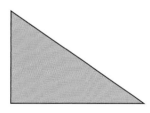

Communication Tip
A shape that can fit on itself only once during one complete rotation has no rotational symmetry, but we say that it has order of rotational symmetry 1.

Checking

3. a) Predict the order of rotational symmetry for this shape.
   b) Determine the order of rotational symmetry for this shape.
   c) Compare the result with your prediction.

Practising

4. a) Trace and cut out the pinwheel.
   b) Predict the order of rotational symmetry for the pinwheel.
   c) Check your prediction by determining the order of rotational symmetry for the pinwheel.

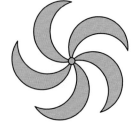

5. a) Predict the order of rotational symmetry for these shapes.
   b) Check your prediction.

6. a) Predict the order of rotational symmetry of these triangles. Explain each prediction.
   b) Check your predictions.
   c) Compare the results with your predictions.

 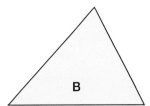 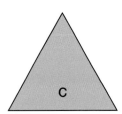

7. a) Sort the shapes into those with rotational symmetry and those without. Use a Venn diagram.
   b) For each shape, list its order of rotational symmetry.

 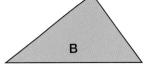

8. a) Can a polygon with no sides of equal length have rotational symmetry? Explain.
   b) Does every polygon with at least two sides of equal length have rotational symmetry? Explain.
   c) Why is a circle the shape with the most rotational symmetry?

Curious Math

Alphabet Symmetry

You will need
- Letters (BLM)

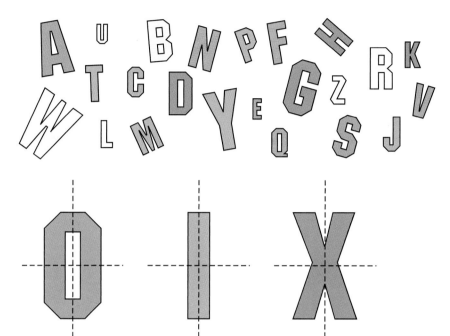

The letters O, I, and X all have two perpendicular lines of symmetry and they also have rotational symmetry.

1 Check the letters of the alphabet. How many lines of symmetry does each letter have?

2 Check each letter for rotational symmetry.

3 For which letters (if any) is each true?
- 0 lines of symmetry, but rotational symmetry
- 1 line of symmetry and rotational symmetry
- 2 lines of symmetry, but no rotational symmetry
- more than 2 lines of symmetry

CHAPTER 14

Frequently Asked Questions

Q: How can you describe a rotation?

A: A rotation is a turn of a shape. Each point in the shape must stay an equal distance from the centre of rotation.

A rotation is described by identifying the centre of rotation, the angle of rotation, and the direction of rotation. The centre of rotation can be on the shape or outside of the shape.

For example, the parallelogram is rotated 90° CCW around point A. The triangle is rotated 120° CW around point B.

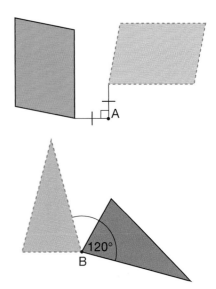

Q: How can a transformation be described in more than one way?

A: Any rotation can be described by the CW or CCW direction.

For example, this trapezoid was rotated 90° once CW or three times CCW about point C.

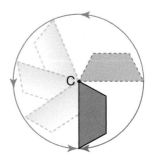

Q: When does a shape have rotational symmetry?

A: A shape has rotational symmetry if it can be turned to fit into its outline more than once. All of the regular polygons have rotational symmetry, but so do some other shapes.

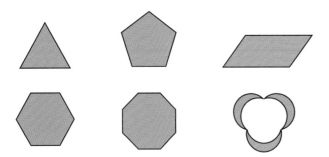

CHAPTER 14

Mid-Chapter Review

LESSON

1
1. a) Describe a rotation that will make this design look different.
   b) Can you describe a rotation that will not change the appearance of this design? Explain.

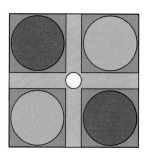

2
2. Frances is playing on a swing.
   a) Describe the swing's movement in both directions as a rotation.
   b) Her friend gave her a push and she's swinging higher. Draw a diagram and describe the new rotation.

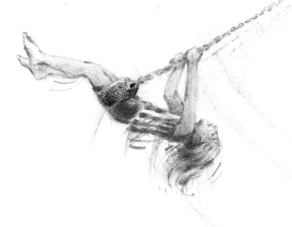

3
3. Determine the order of symmetry for each shape.

CHAPTER 14

4 Communicate Using Diagrams

You will need
- a ruler
- a protractor

Goal Use clear, labelled diagrams to communicate.

Khaled was moving a heavy box. He wanted to move it along the wall, but it was too heavy to push.

? How could you use a diagram to show how to move the box by turning it?

Khaled's Diagram

I'll represent the box with a square. Then I'll show how to rotate the box to move it along the wall.

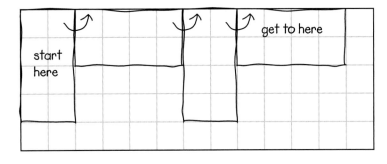

A. What other information should Khaled give in his diagram?

B. Improve Khaled's diagram. Check your directions using the Communication Checklist.

Communication Checklist

☑ Is your diagram easy to understand?

☑ Did you include measurements for all important sides and angles?

☑ Did you use a general enough example or use many examples?

☑ Did you give enough information?

Reflecting

1. How did you decide which other information to include in your diagram?

2. How did you make sure that your diagram made the sequence of the rotations clear?

Checking

3. You want to show that to complete a 90° rotation CW followed by a 180° rotation CCW around the same rotation centre, you could just complete one 90° rotation counterclockwise around that point. Draw a diagram to show why. Use the Communication Checklist.

Practising

4. Use a diagram to show that any parallelogram can be divided into two triangles so that one triangle is the result of a rotation of the other triangle.

5. Draw a diagram to show how to locate a line of reflection using a ruler.

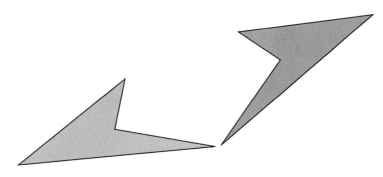

6. Draw diagrams to show that sometimes a double reflection can be described as a translation, and sometimes as a rotation.

7. Draw diagrams to show that there are many times in the day when the hands of a clock show a 90° rotation. Give the times that match at least three of your diagrams.

CHAPTER 14

Exploring Transformation Patterns with Technology

You will need
- GSP software

Goal Relate number patterns to translation, rotation, and reflection patterns.

Ayan noticed a pattern when she translated triangles by moving one vertex to a new position based on doubling both coordinates in its coordinate pair. The first three triangles fell on a line.

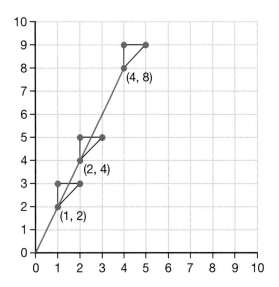

? How can number patterns relate to transformation patterns?

A. Describe the pattern for the first coordinates if the translations continued the same way. What would the 10th term in the pattern be?

B. Describe the pattern for the second coordinates if the translations continued the same way. What would the 10th term in this pattern be?

C. What are the coordinate pairs of the three vertices of the 10th triangle in the pattern? Explain how you know that the tenth triangle lies on the same line as the first three triangles.

D. a) Construct a triangle with vertices at (1, 1), (1, 3) and (2, 2). Rotate it 90° clockwise around (2, 2).
b) Keep rotating 90° clockwise around (2, 2).
c) If each triangle is numbered in order starting at 1, make a list of the triangles that would fall on top of triangle 1.
d) How might you have predicted this pattern?

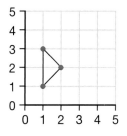

E. a) Select one of the vertices on the letter P. List the coordinates of each of the first three reflections of that point as it moves to the right.
b) Describe the pattern for the first coordinates. What would be the 10th term in the pattern? Why?
c) Describe the pattern for the second coordinates. What would be the 10th term in the pattern? Why?

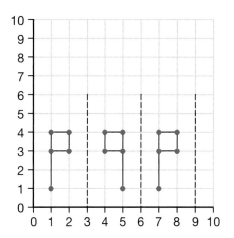

F. a) Create three of your own patterns, one involving translations, one involving rotations, and one involving reflections on a coordinate grid.
b) Draw the visual patterns.
c) Describe the number patterns in the coordinates.

Reflecting

1. Which of your patterns increased by a constant amount?
2. Why did the reflection number patterns in Part E not increase by the same amount each time?
3. How were your translation patterns in Part F similar to the patterns in Parts A and B?
4. How was your rotation pattern in Part F similar to the one in Part D?
5. Which number patterns did you find most predictable? Explain.

CHAPTER 14

6 Creating Designs

You will need
- pattern blocks
- a protractor
- a ruler
- a transparent mirror

Goal: Create a design by performing transformations on a basic shape.

Emilio sketched different designs in his art scrapbook.

? How can you create a design by transforming a shape?

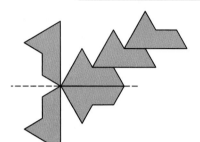

Emilio's Design

I'll use pattern blocks to design a basic shape.

I can rotate the shape about a point.

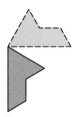

I can also reflect the shape across a line of reflection.

I can also translate the shape.

A. Trace pattern blocks to make your own basic shape.

B. Use the three types of transformation to create a design.

C. Describe how you created the design by performing transformations on your basic shape.

D. Can you transform the basic shape in a different way to create the same design? Describe the transformations.

E. Use the same transformations, but in a different order. How are your designs alike? How are they different?

Reflecting

1. Suppose you reflect a shape and then translate it. Do you expect it to end up in the same place if you did the same transformations in the opposite order? Explain.

2. Which of your transformations moved your shape the farthest from its original position? Why might you have expected that?

3. What strategies did you use to figure out how to create your design in a different way?

Checking

4. a) Create a design by transforming this basic shape. Use all three transformations.
   b) Describe the transformations.
   c) Can you get the same design by doing different transformations?

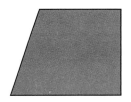

Practising

5. Gen created this basic shape using pattern blocks.
   a) Use all three transformations to create a design using this basic shape. Describe the transformations.
   b) Can you transform the shape in a different way to get the same design? Explain.

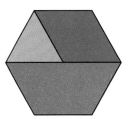

6. a) Describe how this design was created using reflections of one of the basic shapes.
   b) Describe how it can be created using 90° rotations of both basic shapes.
   c) Is it possible to create the same design using 180° rotations? Explain.

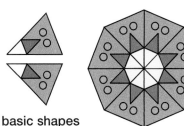

basic shapes

Mental Imagery

Identifying Transformations

A. Which shapes are the result of translating A? Describe the translations.

B. Which shapes are the result of reflecting A? Label the lines of reflection. Check using a transparent mirror.

C. Which shapes are the result of rotating A? Describe the rotations.

D. Which shapes are the result of more than one transformation? Describe the transformations.

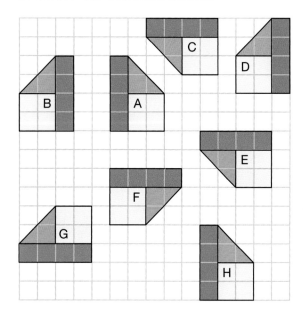

Try These

1. Describe how each shape is the result of one or more transformations of J.

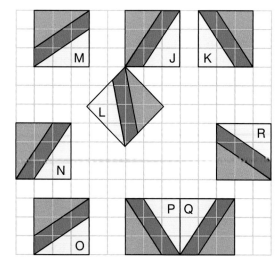

CHAPTER 14

Skills Bank

LESSON

1 1. a) Describe a rotation that will show a different letter.
   b) Choose a different direction and centre of rotation that will also show that letter.

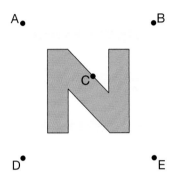

2 2. Describe each rotation. Include the centre of rotation, angle, and direction.

a)

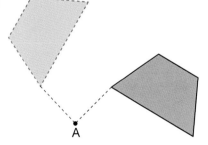

b)

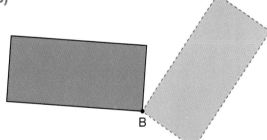

3. Copy the triangle onto grid paper.
   a) Rotate the triangle 90° CW about point C.
   b) Rotate the triangle 180° CCW about point D.

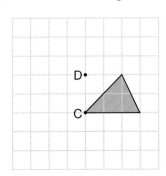

4. Which shapes have rotational symmetry?

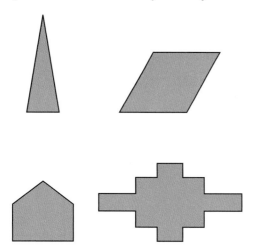

5. Identify the order of rotational symmetry for each shape in Question 4.

6. Create a pattern by translating the triangle repeatedly, with all points moving up 3 and right 2 each time.
   a) Describe the pattern in the first coordinates for the point that started at (1,4).
   b) Describe the pattern in the second coordinates for the same point.
   c) What are the three coordinate pairs for the 15th triangle in the pattern?

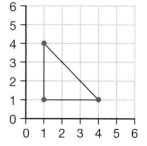

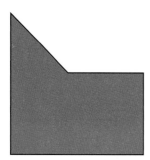

7. a) Create a design by transforming this basic shape. Use all three transformations.
   b) Describe the transformations.

CHAPTER 14

Problem Bank

LESSON

1
1. a) Create a design that looks the same when you rotate it 90° CW or CCW.
   b) Create a design that looks different when you rotate it 90° CW or CCW.

2. Make these turns. Use a die. Describe how you turned the die each time. Continue the pattern and draw the next die.

2
3. It's 2 p.m. The hour hand of the clock moves 90° CW and the minute hand moves 90° CCW.
   a) What time would it be?
   b) Would a clock ever look like this? Explain.

3
4. Sketch a shape that is not a square that has order of rotational symmetry of 4.

5. Sketch a shape with order of rotational symmetry 4, but no lines of symmetry.

6. Draw a square on a coordinate grid with vertices at (3, 4), (4, 5), (4, 3) and (5, 4). Rotate the square repeatedly 90° CW about (3, 4).
   a) Each time you rotate, list the new coordinates for the vertex that began as (4, 5).
   b) Describe the patterns in the first and second coordinates in part a).
   c) Do you get a similar pattern if you follow the vertex at (4, 3) instead? (5, 4)? Explain.

6
7. a) Which of the digits from 0 through 9 look the same when reflected? Describe the reflections.
   b) Which of the digits from 0 through 9 look the same when rotated? Describe the rotations.

CHAPTER 14

Frequently Asked Questions

Q: What are important things to consider when communicating about mathematical ideas using diagrams?

A: It is important that:
- your diagram be easy to understand
- you include all necessary measurements
- you use enough examples to display what you are trying to show
- you provide enough information

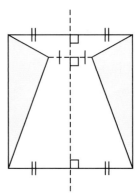

For example, to show that the transformation of the left triangle was a reflection, you should record all necessary measurements.

Q: How can you create designs using transformations?

A: You can begin with one or more shapes. Then you can apply translations, rotations, and reflections in a way that allows the shapes to be put together to make a design.

For example, this shape can be turned into a design by reflecting, rotating, and translating.

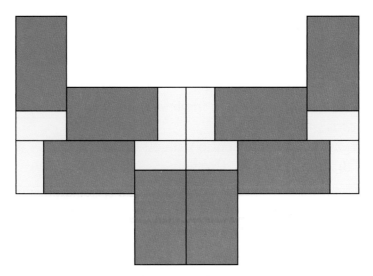

CHAPTER 14

Chapter Review

LESSON

1 1. Rotate the cracker either 90° or 180° and either CW or CCW around each of the three points A, B, and C. When does the cracker end up in the same new position?

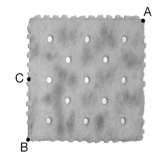

2 2. Musicians use metronomes to keep time when they practise. The arm of the metronome swings at different speeds for different tempos.

Copy the metronome arm. Show where it would be if it were rotated 90° CCW from its present position around the point at the bottom.

3. a) Draw the first 5 terms of the pattern that follows this rule:
    Start with the pentagon at the right.
    Rotate each shape 180° CCW around point A.
   b) Which shapes in the pattern, if you continued it, would look like the third shape?

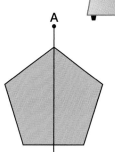

3 4. Which of these shapes has the greatest order of rotational symmetry? Explain.

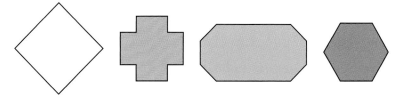

5. Draw a shape for each.
   a) has rotational symmetry, but not line symmetry
   b) has line symmetry, but not rotational symmetry

6 6. a) Create a basic shape using several pattern blocks.
   b) Use all three transformations to create a design using this basic shape. Describe the transformations.
   c) Can you transform the shape in a different way to get the same design? Explain.

7. Describe a transformation on the letter b that will result in each letter.
   a) d   b) p   c) q

CHAPTER 14

Chapter Task

Creating Nets

Maggie is drawing a poster showing all of the nets of a cube. She notices that she could describe each net using transformations.

For example, to get from the pink square to the blue one, she could rotate the pink square 90° CW around its bottom right vertex.

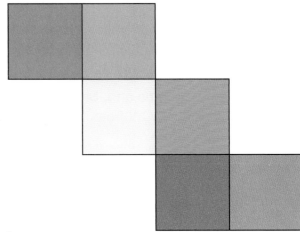

? How can you describe how to create each net for a cube using transformations?

A. Describe how to complete the net shown starting with one square and using only translations, only reflections, or only rotations.

B. Describe how to complete the net shown starting with one square and using a combination of transformations.

C. Create at least three other nets for cubes. For each one, try to describe how to create it using only one kind of transformation. Then try to describe how to create it using all three transformations.

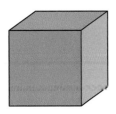

Task Checklist

☑ Are your descriptions clear and complete?

☑ Did you use the correct number of transformations each time?

☑ Are your diagrams clear and labelled?

Cumulative Review

CHAPTERS 12–14

Cross-Strand Multiple Choice

1. Which ■ can be replaced by >?

   A. $\frac{3}{4}$ ■ $\frac{4}{5}$ B. $1\frac{5}{6}$ ■ $\frac{5}{3}$ C. $1\frac{2}{5}$ ■ $\frac{7}{4}$ D. $\frac{7}{8}$ ■ $1\frac{3}{4}$

2. Which is a decimal equivalent for $\frac{4}{5}$?

   A. 0.08   B. 0.40   C. 0.8   D. 0.45

3. At a dock, 3 of the 14 boats are sailboats. The others are motorboats. Which is the ratio of motorboats to sailboats?

   A. 14 : 3   B. 3 : 11   C. 3 : 14   D. 11 : 3

4. Water covers about $\frac{3}{4}$ of the world's surface. Which percent is this?

   A. 34%   B. 25%   C. 75%   D. 50%

5. Which is closest to the percent of the snowboard that is red?

   A. 100%   B. 50%   C. 25%   D. 10%

6. Qi bought 3 videos on sale for $43.50. Which is the unit rate?

   A. $14.50 for 1 video
   B. $130.50 for 1 video
   C. $21.75 for 1 video
   D. $7.25 for 1 video

7. Tara is going to spin the spinner 50 times. Which is the best prediction for the number of times she will spin purple?

   A. 20   B. 25   C. 15   D. 10

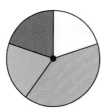

8. Akeem is spinning this spinner twice. Which is the number of possible outcomes?

   A. 16   B. 4   C. 12   D. 9

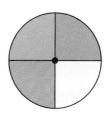

9. Raven is rolling two regular dice. Which is the theoretical probability that the product will be greater than 14?

   A. $\frac{23}{36}$ B. $\frac{20}{36}$ C. $\frac{13}{36}$ D. $\frac{12}{36}$

10. Denise rolled a die 20 times. Which result is not possible?

    A. $\frac{4}{5}$ of the rolls were even numbers

    B. $\frac{6}{5}$ of the rolls were greater than 1

    C. $\frac{1}{2}$ of the rolls were multiples of 3

    D. $\frac{1}{10}$ of the rolls were factors of 4

11. Which is the next term?

    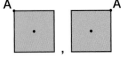 , , , , ...

    A.

    C.

    B.

    D.

12. Which rotation about the centre of the orange circle will make the design look the same?

    A. 90° CW   C. 90° CCW
    B. 180° CW   D. none of these

13. The green polygon is rotated onto the yellow polygon. Which describes the rotation?
    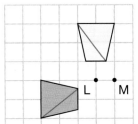
    A. 90° CW about point M
    B. 90° CW about point L
    C. 90° CCW about point L
    D. 90° CCW about point M

Cross-Strand Investigation

Music and Decorations for a Class Party

The students are planning a class party for the end of the school year. Raven, Marc, and Isabella are on the decoration and music committee.

14. To decide the order that songs will be played, Raven writes "rock", "pop", and "hip hop" on three slips of paper and puts the slips in a bag. She draws a slip, records the result, and returns the slip to the bag. Then she draws again.
    a) Use a tree diagram to list the possible outcomes for the first two draws.
    b) What is the probability the first two songs will be the same kind of music?

15. Marc is buying supplies and making decorations.
    a) Which colour of ribbon is the best bargain?
    b) Marc mixed 4 L of green paint with 3 L of white paint. Write two part to whole ratios about the paint. Tell what each ratio represents.
    c) Marc made a mural about the class year. He used $1\frac{2}{5}$ packages of purple paper, $\frac{3}{4}$ of a package of white paper, and $\frac{4}{3}$ packages of blue paper. All of the packages are the same size. Which colour did he use most? Which did he use least?
    d) Create a design for a wall decoration for the class party by transforming this basic shape. Use all three transformations.
    e) Describe the transformations in your design for Part d).

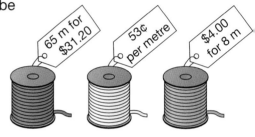

16. a) Isabella is making paper mats for party snacks. Each mat will be a triangle or a parallelogram. The base of one triangle mat is 22 cm, and the height is 18 cm. What is the area of the mat?
    b) Each mat will have the same area. What height and base can Isabella use for a parallelogram mat? Give three answers.

Party Games

Maggie and Jorge are organizing the games. They sorted the game suggestions and made a tally chart.

Kinds of Games	
Guessing	|||
Dice	||||
Board	||||	
Bean bag		
Spinner	||||	
Ball	||

17. a) What is the ratio of dice games to bean bag games?
    b) List the percents for each kind of game.
    c) In one of the dice games, Nine Prime, the player rolls a die. If the number is a factor of 9, the player scores two points. If the number is a prime number, the player scores one point. What is the theoretical probability of each event?
    d) Conduct an experiment for Nine Prime with 24 rolls. What is your experimental probability for each event?
    e) Suppose you repeated the experiment from part e) and combined the data. Do you think the experimental probability would be closer to the theoretical probability? Explain.
    f) Maggie made this spinner for one of the spinner games. Describe a rotation that will take the spinner arrow to the line between the blue and green sectors.
    g) Describe two different rotations for the spinner arrow to go to the line between the purple and green sectors.

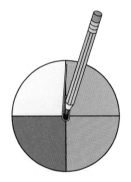

18. a) When playing Double Spin, a player spins the spinner twice. The sum is the player's score for that turn. What is the greatest possible score in one turn? What is the least possible score?
    b) Maggie is constructing a parallelogram with sides 8 cm and 4 cm for the centre of a gameboard. The angle measures are 75° and 105°. Draw the parallelogram.
    c) Jorge has 200 cm³ of modelling clay to make game pieces. Each game piece will be a rectangular prism 4 cm high with a square base with 1 cm sides. How many game pieces can he make?
    d) Jorge is going to paint all the sides of the game pieces he makes. What is the total area he will paint?
    e) Maggie brought this box of bean bags for a game. What is the mass of each bean bag?

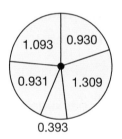

Glossary

Instructional Words

calculate: Figure out the number that answers a question; compute

clarify: Make a statement easier to understand; provide an example

classify: Put things into groups according to a rule and label the groups; organize into categories

compare: Look at two or more objects or numbers and identify how they are the same and how they are different (e.g., Compare the numbers 6.5 and 5.6. Compare two shapes.)

construct: Make or build a model; draw an accurate geometric shape (e.g., Use a ruler and a protractor to construct an angle.)

create: Make your own example

describe: Tell, draw, or write about what something is or what something looks like; tell about a process in a step-by-step way

draw: 1. Show something in picture form (e.g., Draw a diagram.)
2. Pull or select an object (e.g., Draw a card from the deck. Draw a tile from the bag.)

estimate: Use your knowledge to make a sensible decision about an amount; make a reasonable guess (e.g., Estimate how long it takes to cycle from your home to school. Estimate how many leaves are on a tree. Estimate the sum of 3210 and 789.)

evaluate: Determine if something makes sense; judge

explain: Tell what you did; show your mathematical thinking at every stage; show how you know

explore: Investigate a problem by questioning, brainstorming, and trying new ideas

extend: 1. In patterning, continue the pattern
2. In problem solving, create a new problem that takes the idea of the original problem further

justify: Give convincing reasons for a prediction, an estimate, or a solution; tell why you think your answer is correct

list: Record thoughts or things one under the other

measure: Use a tool to describe an object or determine an amount (e.g., Use a ruler to measure the height or distance around something. Use a protractor to measure an angle. Use balance scales to measure mass. Use a measuring cup to measure capacity. Use a stopwatch to measure the time in seconds or minutes.)

model: Show an idea using objects and/or pictures (e.g., Model a number using base ten blocks.)

predict: Use what you know to work out what is going to happen (e.g., Predict the next number in the pattern 1, 2, 4, 8, ….)

reason: Develop ideas and relate them to the purpose of the task and to each other; analyze relevant information to show understanding

record: Put work in writing or in pictures

relate: Show a connection between objects, drawings, ideas, or numbers

represent: Show information or an idea in a different way (e.g., Draw a graph. Make a model. Create a rhyme.)

show (your work): Record all calculations, drawings, numbers, words, or symbols that make up the solution

sketch: Make a rough drawing (e.g., Sketch a picture of the field with dimensions.)

solve: Develop and carry out a process for finding a solution to a problem

sort: Separate a set of objects, drawings, ideas, or numbers according to an attribute (e.g., Sort 2-D shapes by the number of sides.)

validate: Check an idea by showing that it works

verify: Work out an answer or solution again, usually in another way, to show that the original answer is correct; show evidence of correctness

visualize: Form a picture in your mind of what something is like; imagine

Glossary

Mathematical Words

acute angle: An angle that measures less than 90°

acute triangle: A triangle with only acute angles

addend: A number that is added to another number

algorithm: A series of steps you can use to carry out a procedure (e.g., add, subtract, multiply, or divide)

analog clock: A clock that measures time using rotating hands

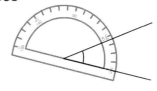

angle: An amount of turn measured in **degrees**

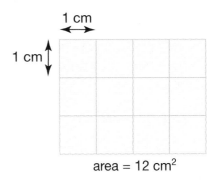

area: The amount of space a surface covers, measured in square units

area = 12 cm²

array: A rectangular arrangement of objects or pictures in **rows** and **columns** (e.g., An array can show why 2 × 3 and 3 × 2 have the same product.)

This array shows 2 rows of 3 or 2 × 3.
It also shows 3 columns of 2 or 3 × 2.
In both cases, the product is 6.

attribute: A characteristic or quality that can be used to describe and compare things (e.g., Some common attributes of shapes are size, colour, texture, and number of edges.)

average: One piece of data that is a good overall representative of all of the pieces of data in a set; there are different types of averages. (See also **mean** and **mode**.)

axis (plural is axes): A horizontal or vertical line in a graph, labelled with words or numbers to show what the lines, bars, or pictures in the graph mean

B

bar graph: A way to show and compare data that uses horizontal or vertical bars

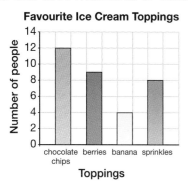

base: 1. The **face** on which a 3-D shape is resting
2. The face that determines the name and the number of edges of a **prism** or **pyramid**
3. The **line segment** at the bottom of a 2–D shape

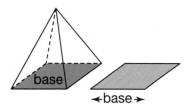

base ten blocks: Blocks that represent numbers as hundredths, tenths, ones, tens, hundreds, thousands, and so on

Thousands	Hundreds	Tens	Ones
2	3	5	4

biased results: Survey results for part of a group that are not likely to apply to the rest of the group

broken-line graph: A graph in which data points are connected point to point

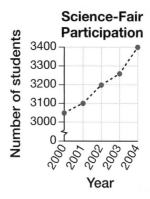

C

capacity: The amount that a container will hold; common units of measurement are **litres (L)** and **millilitres (mL)**

Carroll diagram: A diagram that uses rows and columns to show relationships

Sorting Whole Numbers From 1 to 10

	Greater than 5	Less than or equal to 5
Even	6, 8, 10	2, 4
Odd	7, 9	1, 3, 5

cell: A box in a spreadsheet or table (e.g., Cell A4 is in column A and row 4.)

centimetre (cm): A unit of measurement for **length**; one hundredth of a metre (e.g., A fingertip is about 1 cm wide.)
1 cm = 10 mm, 100 cm = 1 m

centre of rotation: The point that a shape rotates around (e.g., Point O is the centre of rotation for the triangle.)

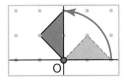

century: A unit of measurement for time; 100 years

certain outcome: A result that will always occur (e.g., If you roll a die with a 3 on every face, rolling a 3 is a certain outcome.)

chance: The likelihood that a particular event will occur. (See also **probability**.)

circle: The curve formed by a set of points that are all the same distance from the centre

circle graph: A way to show data that uses parts of a circle to represent parts of the set of data (e.g., A circle graph can be used to show how students spend their days.)

circumference: The distance around a **circle**

clockwise (CW): The direction the hands of an **analog clock** move

closed: Having no **endpoints** (e.g., A square is a closed shape.)

column: A set of items lined up vertically (See also **row**.)

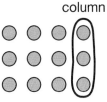

common difference: The difference between any two terms in a pattern, if it is always the same for each pair of terms (e.g., In the pattern 3, 7, 11, 15, … the common difference is 4.)

composite number: A number that has more than two different factors (e.g., 4 is a composite number because it has more than 2 factors: 1, 2, and 4.)

concave: Curved or pointed inward (e.g., A concave **polygon** has one **vertex** that points inward.)

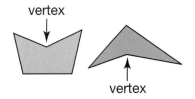

congruent: Shapes in which all matching angles and sides are equal

These shapes are congruent.

convex: Curved or pointed outward (e.g., A convex **polygon** has all of its **vertices** pointing outward.)

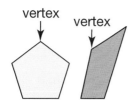

coordinate grid: A grid with horizontal and vertical lines numbered in order

coordinate pair: A pair of numbers that describes a point where a **vertical** and a **horizontal** line meet on a coordinate grid; the coordinate from the horizontal axis is always written first (e.g., Point A has the coordinates (3, 4).)

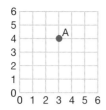

counterclockwise (CCW): The opposite direction from **clockwise**

cube: A 3-D shape with six **congruent** square faces

cubic centimetre (cm³): A unit of measurement for **volume**; the volume occupied by a cube with edges all 1 cm

data: Information gathered in a survey, in an experiment, or by observing (e.g., Data can be in words, such as a list of students' names; in numbers, such as quiz marks; or in pictures, such as drawings of favourite pets.)

decade: A unit of measurement for time; 10 years

decametre: A unit of measurement for length 10 m (e.g., A classroom is about 1 dam long.)
10 m = 1 dam, 100 dam = 1 km

decimal: A way to describe fractions and mixed numbers using place value; a **decimal point** separates the ones place from the tenths place

decimal equivalent: A decimal that represents the same part of a whole or the same part of a set as a fraction or another decimal
$\frac{5}{10} = 0.5$ and $\frac{5}{100} = 0.05$ and $0.5 = 0.50$

decimal point: A dot used to separate the whole number part from the fractional part in a decimal

decimetre (dm): A unit of measurement for **length**; one tenth of a **metre** (e.g., A tens rod from a set of **base ten blocks** is 1 dm long.)
1 dm = 10 cm, 10 dm = 1 m

degree (°): A unit of measurement for angle size

This angle is 90°.

degree Celsius (°C): A unit of measurement for temperature (e.g., Water freezes at 0°C and boils at 100°C.)

denominator: The number below the bar in a **fraction** symbol that represents the number of parts in the whole or set. (See also **numerator**.) (e.g., The denominator of $\frac{3}{4}$ is 4.)

$\frac{3}{4}$ ← denominator

diagonal: In a 2-D shape, a line segment that joins any two **vertices** that are not next to each other; in a 3-D shape, a diagonal joins any two vertices that are not on the same **face**

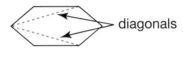

difference: The result when you subtract; the amount by which one number is greater than or less than another number

$$\begin{array}{r} 93 \\ -45 \\ \hline 48 \end{array}$$ ← difference

digit: One of the ten number symbols from 0 to 9; in the base-ten system, the position of a digit shows its value (e.g., The digit 3 in the number 237 represents 3 tens; in 5.03, it represents 3 hundredths.)

dimensions: The measures of an object that can be used to describe its size (e.g., The dimension of a line segment is length; the dimensions of a rectangle are length and width; the dimensions of a rectangular prism are length, width, and height.)

dividend: The number that is divided into equal parts in a division operation

9 ÷ 3 = 3
↑
dividend

divisible: Can be divided with no remainder (e.g., 30 is divisible by 6 because you can make exactly 5 groups of 6 from 30.)

divisor: The number you divide by in a division operation

24 ÷ 3 = 8
↑
divisor

double: To multiply a number by 2

E

edge: The **line segment** formed where two **faces** meet on a 3-D shape

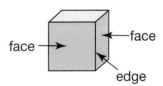

endpoint: The point at which a **line segment** begins or ends

equally likely outcomes: Results that have an equal chance of occurring (e.g., In flipping a coin, heads and tails are equally likely outcomes.)

equation: A mathematical statement in which the left side is equal to the right side

4 + 2 = 6
5 + 3 = 4 + 4

equilateral: In a triangle, having all sides equal in length

equivalent: Having the same value (e.g., Equivalent fractions are two fractions that represent the same number, such as $\frac{2}{6}$ and $\frac{1}{3}$.)

equivalent decimal: See **decimal equivalent**

equivalent fractions: Fractions that represent the same part of a whole or the same part of a set

$\frac{2}{4}$ is equivalent to $\frac{1}{2}$.

$$\frac{2}{4} = \frac{1}{2}$$

equivalent ratios: Two or more ratios that represent the same comparison

estimate: A reasoned guess about a measurement or answer

even number: A number that is **divisible** by 2 (e.g., 12 is even because 12 ÷ 2 = 6.)

event: A set of **possible outcomes** of a **probability** experiment (e.g., When rolling a die, you could decide that an event is rolling an even number, such as 2, 4, or 6.)

expanded form: A way to write a number that shows the value of each digit (e.g., In expanded form, 2365 is 2000 + 300 + 60 + 5 or 2 thousands 3 hundreds 6 tens 5 ones.)

expression: A mathematical statement made with numbers or variables and symbols (e.g., 5 + 3 − 7 is an expression.)

face: A 2-D shape that forms a flat surface of a 3-D shape

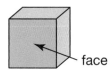

face

fact: An addition, subtraction, multiplication, or division **equation** (e.g., In Grade 5, we learn multiplication facts to 12 × 12 = 144.)

fact

fact family: A set of addition and subtraction or multiplication and division facts; all facts in the set use the same numbers

$$3 \times 2 = 6 \quad 6 \div 3 = 2$$
$$2 \times 3 = 6 \quad 6 \div 2 = 3$$

factor: A whole number that divides another whole number without a remainder (e.g., 8 ÷ 2 = 4; 2 is a factor of 8 because 2 divides 8 without a remainder.)

fair game: A game that all players have an equal chance of winning

flip: See **reflection**. (See also **transformation**.)

fraction: Numbers used to name part of a whole or part of a set
(e.g., $\frac{3}{4}$ is a **proper fraction**;
$\frac{4}{3}$ is an **improper fraction**;
0.2 is a **decimal** fraction;
$5\frac{1}{2}$ is a **mixed number**.)
(See also **numerator** and **denominator**.)

frequency: The number of times an event occurs

gram (g): A unit of measurement for **mass** (e.g., 1 mL of water has a mass of 1 g.) 1000 g = 1 kg

graph: A way of showing information so it is more easily understood. A graph can be concrete (e.g., boys in one line and girls in another), pictorial (e.g., pictures of boys in one row and girls in another), or abstract (e.g., two bars on a bar graph to show how many students are boys and how many are girls)

greater than (>): A sign used when comparing two numbers (e.g., 10 is greater than 5, or 10 > 5.)

h: The symbol for **hour**

halve: To divide a number by 2

height: The distance from a side of a polygon to an opposite vertex, measured using a line segment perpendicular to the side

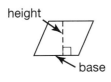

heptagon: A **polygon** with seven sides

hexagon: A **polygon** with six sides

horizontal line: A line across a page that is **parallel** to the bottom edge, or a line that is level with the floor

hour: A unit of measurement for time; the symbol for hour is **h**
60 min = 1 h

hypothesis: A statement that you think you can test

impossible outcome: A result that cannot occur (e.g., If you roll a die with a 3 on every face, rolling a 5 is an impossible outcome.)

improper fraction: A fraction in which the **numerator** is greater than the **denominator** $\left(\text{e.g., } \frac{4}{3}\right)$

inequality signs: The symbols > and < that are used to make comparisons (e.g., 8 > 5 is read as "Eight is greater than five." 5 < 8 is read as "Five is less than eight.") (See also **greater than** and **less than**.)

interval: The distance between two endpoints on a graph scale; intervals in a graph should be equal (e.g., If the scale axis is numbered 0, 5, 10, 15, …, then the intervals are 5.)

isosceles: In a triangle, having two sides equal in length

isometric sketch: A 3-D view of an object on isometric dot paper. All equal lengths on the cubes are equal on the grid

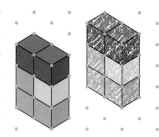

kilogram (kg): A unit of measurement for **mass** (e.g., A math textbook has a mass of about 1 kg.)
1 kg = 1000 g

kilometre (km): A unit of measurement for **length**; one thousand metres
1 km = 1000 m

kite: A quadrilateral that has two pairs of equal sides with no sides parallel

legend: A feature on a map or graph that explains what colours or symbols mean

length: The distance from one end of a **line segment** to the other end

The length of this line segment is 2 cm.

less than (<): A sign used when comparing two numbers (e.g., 5 is less than 10, or 5 < 10.)

like denominators: Equal denominators (e.g., $\frac{3}{8}$ and $\frac{7}{8}$ are both in eighths; they have like denominators.)

likely outcome: A result that can easily occur (e.g., If you roll a die with a 3 on all the faces but one, rolling a 3 is a likely outcome.)

line graph: A graph of a smooth line through points that shows how the change in one value is related to change in another value

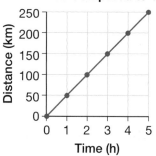

line of symmetry: A line that divides a 2-D shape into halves that match when the shape is folded on the line of symmetry

line of symmetry

linear pattern: A pattern in which the difference between each item and the next is always the same (e.g., 2, 4, 6, 8, 10, ... is linear because it always increases by 2. 98, 96, 94, 92, 90, ... is linear because it always decreases by 2.)

line segment: Part of a line with two **endpoints**

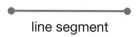

line segment

litre (L): A unit of measurement for **capacity**
1 L = 1000 mL

logical reasoning: A process for using the information you have to reach a conclusion (e.g., If you know all the students in a class like ice cream and that Jane is in the class, you can logically reason that Jane likes ice cream.)

mass: The amount of matter in an object; common units of measurement are grams (g) and kilograms (kg)

mean: A typical value for a set of numbers, determined by calculating the sum of the numbers and dividing by the number of numbers in the set (e.g., The set 3, 4, 5, 2, 2, 3, and 2 has seven numbers with a sum of 21. The mean is 21 ÷ 7 = 3.)

median: A typical value for a set of numbers, determined by ordering the numbers and identifying the middle number (e.g., The set 3, 4, 5, 2, 2, 3, and 2 can be ordered as 2, 2, 2, 3, 3, 4, 5. The median is 3.)

metre (m): A unit of measurement for **length** (e.g., 1 m is about the distance from a doorknob to the floor.)
1000 mm = 1 m, 100 cm = 1 m, 1000 m = 1 km

millennium: A unit of measurement for time; 1000 years

millilitre (mL): A unit of measurement for **capacity**; 1000 mL = 1 L

million: The number that is 1000 thousands; 1 000 000

millimetre (mm): A unit of measurement for **length** (e.g., A dime is about 1 mm thick.)
10 mm = 1 cm, 1000 mm = 1 m

min: The symbol for **minute**

minute: A unit of measurement for time; the symbol for minute is min
60 s = 1 min, 60 min = 1 h

mixed number: A number made up of a **whole number** and a **fraction** (e.g., $3\frac{1}{2}$)

mode: A typical value for a set of numbers, determined by identifying the number that occurs most often in the set (e.g., In the set 3, 4, 5, 2, 2, 3, and 2, the most frequent value is 2. The mode is 2.)

multiple: A number that is the product of two factors (e.g., 8 is a multiple of 2 because 2 × 4 = 8.)

multiples: The products found by multiplying a whole number by other whole numbers (e.g., When you multiply 10 by the whole numbers 0 to 4, you get the multiples 0, 10, 20, 30, and 40.)

nearest (unit): The closest when rounding a number or a measurement; less than half a unit is rounded down to a lesser value; more than half a unit is rounded up to a greater value (e.g., 4.6 cm² rounded to the nearest square centimetre is 5 cm².)

net: A 2-D shape you can fold to create a 3-D shape

This is a net for a cube.

nonlinear pattern: A pattern in which the difference between each term and the next does not stay the same (e.g., 1, 3, 6, 10, 15, ... is nonlinear because the differences are 2, 3, 4, and so on.)

nonstandard unit: A unit of measurement that is not part of a customary system (e.g., A desk is about 5 juice cans wide.)

number line: A diagram that shows ordered numbers or points on a line

number sentence: A mathematical statement that shows how two quantities are related (e.g., 3 × 8 = 24; 3 < 8) (See also **equation**.)

numeral: The written symbol for a number (e.g., 148, $\frac{3}{4}$, and 2.8)

numerator: The number above the bar in a **fraction** symbol. (See also **denominator**.)

$\frac{3}{4}$ ← numerator

obtuse angle: An angle that measures greater than 90° and less than 180°

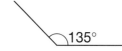

obtuse triangle: A triangle with one obtuse angle

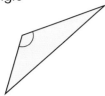

octagon: A polygon with eight sides

odd number: A number that has a remainder of 1 when it is divided by 2 (e.g., 15 is odd because 15 ÷ 2 = 7 R1.)

open sentence: A number sentence containing at least one unknown number (e.g., 2 × ■ = 8)

opposite side: In a triangle, the side an angle does not touch (e.g., the opposite side of ∠Q is q.)

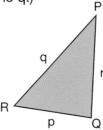

order of rotational symmetry: The number of times a shape will fit on itself exactly during one complete rotation (e.g., A square has order of rotational symmetry of 4.)

ordinal number: A way of describing an item's place in a numbered sequence (e.g., 1st, third, 15th)

organized list: The problem-solving strategy of following an order to find all possibilities

orientation: The direction around a shape when you name the vertices in order

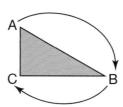 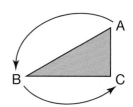

triangle ABC with clockwise orientation

triangle ABC with counterclockwise orientation

origin: The point on a coordinate grid at which the horizontal and vertical axes meet (E.g., The coordinate pair of the origin is (0, 0).)

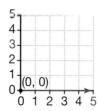

outcome: A single result (e.g., If you roll a die, the possible outcomes are 1, 2, 3, 4, 5, and 6; 7 is an impossible outcome.)

parallel: Always the same distance apart

parallelogram: A **quadrilateral** with equal and **parallel** opposite sides (e.g., A **rhombus**, a **rectangle**, and a **square** are all types of parallelograms.)

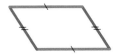

pattern: Something that follows a rule while repeating or changing

pattern rule, explicit: A pattern rule that uses the term number (*n*) to determine a term in the pattern (e.g., "4 + 2(*n* − 1)" is an explicit pattern rule for 4, 6, 8, 10, ….)

pattern rule, recursive: A pattern rule that gives the first term of a pattern and tells how the pattern continues (e.g., "Start with 4 and add 2 each time" is a recursive pattern rule for 4, 6, 8, 10, ….)

pentagon: A **polygon** with five sides

percent: A part-to-whole ratio that compares a number or an amount to 100
$$\left(\text{e.g., } 25\% = 25:100 = \frac{25}{100}\right)$$

perimeter: The total length of the sides of a shape

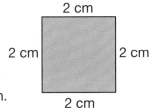

The perimeter of this square is 8 cm.

perpendicular: At right angles (e.g., The line segment AB is perpendicular to CD.)

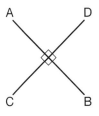

pictograph: A **graph** that uses pictures or symbols to represent quantities

How Old Are You?

7 ☺☺☺☺
8 ☺☺☺☺☺☺☺☺
9 ☺☺☺☺☺☺☺
10 ☺☺☺☺

Each ☺ means 5 people.

place value: The value given to a digit based on its position in a numeral (e.g., The 3 in the number 237 represents 3 tens, while in the number 5.03 it represents 3 hundredths.)

plot: Locate and draw a point on a coordinate grid

polygon: A closed 2-D shape with sides made from straight line segments

polyhedron: A 3-D shape with polygons as faces (e.g., Prisms and pyramids are two kinds of polyhedron.)

possible outcome: Any result that can occur (e.g., If you roll a die, the possible outcomes are 1, 2, 3, 4, 5, and 6.)

precision: A way to compare tools and measurements (e.g., A measurement made with a ruler divided in millimetres is more precise than a measurement made with a ruler divided in centimetres.)

prime number: A number that has only two different factors: 1 and itself (e.g., 2 is a prime number because it has only two factors: 1 and 2.)

prism: A 3-D shape with opposite **congruent bases**; the other faces are parallelograms (e.g., a triangular prism)

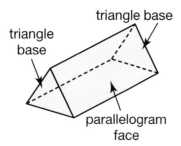

probability, experimental: The probability that results from an experiment (e.g., If you flip a coin 20 times and 11 of those flips are a head, then the experimental probability of flipping a head is $\frac{11}{20}$.)

probability, theoretical: The probability you would expect when you analyze all of the different possible outcomes (e.g., The theoretical probability of flipping a head on a coin is $\frac{1}{2}$, since there are 2 equally likely outcomes and only 1 is a head.)

probability line: A way to show probabilities of several outcomes

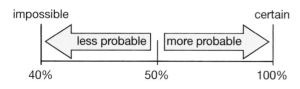

product: The result when you multiply

2 × 6 = 12
         ↑
      product

proper fraction: A fraction in which the **denominator** is greater than the **numerator** (e.g., $\frac{1}{2}, \frac{5}{6}, \frac{2}{7}$)

properties: The features of a shape that describe it (e.g., The properties of a square are four equal sides, four equal angles (all right angles), two pairs of parallel sides, and four lines of symmetry.)

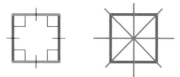

protractor: A tool used to measure **angles**

pyramid: A 3-D shape with a polygon for a base; the other faces are triangles that meet at a single **vertex** (e.g., a rectangular pyramid)

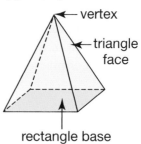

Q

quadrilateral: A polygon with four straight sides. (See also **kite, parallelogram, rectangle, rhombus, square, trapezoid**.)

quotient: The result when you divide, not including the **remainder**

12 ÷ 5 = 2 R2
            ↑
        quotient

random: A result is random if what happens is based on chance; something that is not random has to happen a certain way (e.g., The day after Tuesday is always Wednesday. That's not random. If you put the names of the days of the week in a bag and pick Tuesday out, it is random which day you will pick next.)

range: The **difference** between the greatest and least values in a set of data (e.g., For the numbers 1, 2, 5, 7, 9, 11, 12, the range is 12 − 1 or 11.)

ratio: A comparison of two numbers or quantities measured in the same units. If you mix juice using 1 can of concentrate and 3 cans of water, the ratio of concentrate to water is 1 : 3, or 1 to 3.

rectangle: A **quadrilateral** with four right angles

reflection: A flip of a 2-D shape; each point in a 2-D shape flips to the opposite side of the line of reflection, but stays the same distance from the line. (See also **transformation**.)

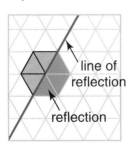

regroup: Trade 10 smaller units for 1 larger unit, or 1 larger unit for 10 smaller units

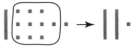

regular polygon: A polygon with all sides equal and all angles equal

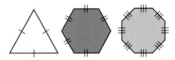

remainder: The number of items left over after division

$$14 \div 4 = 3\ R2$$

remainder

rhombus: A **quadrilateral** with four equal sides

right angle: The angle made by a square corner; a right angle measures 90°

right triangle: A triangle with one right angle

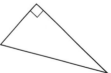

rotation: A turn of a shape; each point in the shape must stay an equal distance from the **centre of rotation**. (See also **transformation**.)

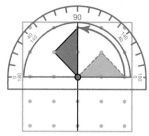

This is a 90° counterclockwise rotation about the centre of rotation.

rotational symmetry: A shape that can fit on itself exactly more than once in one complete rotation has rotational symmetry

round: To approximate a number to a given place value (e.g., 8327 rounded to the nearest hundred is 8300.)

row: A set of items lined up horizontally (See also **column**.)

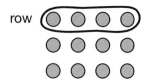

s: The symbol for **second**

scale: 1. Numbers and marks arranged at regular intervals that are used for measurement or to establish position (e.g., the markings on the side of a measuring cup, or on the **axis** of a graph)
2. The size of a model compared with what it represents (e.g., If the scale of a model is "1 cm represents 1 m," a real object that is 1 m tall would be 1 cm tall in the model.)

scale model: A model that is larger or smaller than the real object, but is the same shape; a model is **similar** to the real object

scalene: In a triangle, having no two sides equal in length

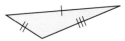

scatter plot: A graph made by plotting coordinate pairs to show if one set of data can be used to make predictions about another set of data

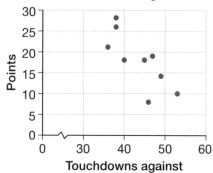

second: A unit of measurement for time; the symbol for second is **s**
60 s = 1 min

set: A collection of items or numbers; each item in the set is called a "member" of the set

shape: 1. A geometric object (e.g., A square is a 2-D shape. A cube is a 3-D shape.)
2. The **attribute** that describes the form of a geometric object (e.g., Circles and spheres both have a round shape.)

side: One of the line segments that forms a polygon

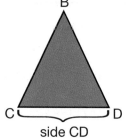

side CD

similar: Shapes in which matching angles are equal and sides have lengths that are all the same multiple of the matching sides (e.g., These two triangles are similar.)

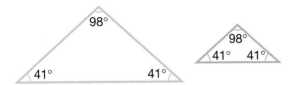

skeleton: A 3-D shape that has only edges and vertices

skip count: To count without using every number, but according to a set pattern or rule (e.g., counting to 100 by fives)

slide: See **translation**. (See also **transformation**.)

square: A **quadrilateral** with four equal sides and four right angles

square centimetre (cm²): A unit of measurement for **area**; the area covered by a square with sides all 1 cm

square corner: See **right angle**

square metre (m²): A unit of measurement for **area**; the area covered by a square with sides all 1 m

square unit: A unit of measurement for **area**

standard form: The usual way in which we write numbers (e.g., 23 650 is written in standard form.) (See also **expanded form** and **numeral**.)

standard unit: A unit of measurement that is part of an accepted measurement system (e.g., Metres, kilograms, litres, and square metres are all standard units.) (See also **nonstandard unit**.)

stem-and-leaf plot: A way to organize data in groups according to place value; The stem shows the beginning of a number and the leaf shows the rest of the number (e.g., The circled leaf in this stem-and-leaf plot represents the number 258.)

Stem	Leaves
24	1 5 8
25	2 2 3 4 7 ⑧ 9
26	0 3
27	
28	8

straight angle: an angle of exactly 180°

180°
―――•―――
A
∠A = 180°

sum: The result when you add

14 + 37 = 51
             ↑
             sum

surface area: The total area of all of the faces, or surfaces, of a polyhedron (e.g., The surface area of this cube is 24 cm² because each face has an area of 4 cm².)

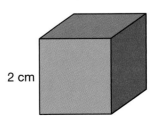

2 cm

survey: 1. A set of questions designed to obtain information directly from people 2. To ask a group of people a set of questions

2-D shape: A shape that has the dimensions of length and width

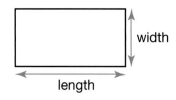

3-D shape: A shape that has the dimensions of length, width, and height

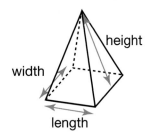

12 h clock: A method of naming the hours of the day from 1 to 12, along with the notation a.m. and p.m.; The symbol for **hour** is **h** (e.g., 2:00 p.m. is equivalent to 14:00.)

24 h clock: A method of naming the hours of the day from 0 to 23; (e.g., 14:00 is equivalent to 2:00 p.m.)

tally: A way to keep track of data using marks

HH HH /

tally chart: A chart that uses tally marks to count data (e.g., If you are surveying students about favourite flavours of ice cream you could use a tally chart like this one.)

Favourite Ice Cream Flavours	
vanilla	HH HH /
chocolate	HH HH HH HH ///
strawberry	HH

t-chart: A way to organize related information in a chart with two columns; both sides of the t are labelled

Weeks	Days
1	7
2	14
3	21

term number: A number that tells the position of a term in a pattern (e.g., In the pattern 1, 3, 5, 7, … the third term is 5.)

tetrahedron: A 3-D shape with four **faces** that are **polygons**

A triangle-based pyramid is a tetrahedron.

tile: Use repeated congruent shapes to cover an area without gaps or overlaps

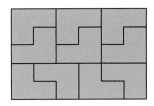

tiling pattern: A pattern of repeated congruent shapes that fit together with no gaps or overlaps

tonne (t): A unit of measurement for mass (e.g., A small car has a mass of about 1 t.) 1 t = 1000 kg

transformation: A rule that results in a change of position, orientation, or size; transformations include **translations**, **rotations**, and **reflections**

translation: A slide of a shape; the slide must be along a straight line. (See also **transformation**.)

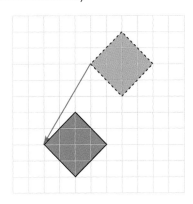

trapezoid: A **quadrilateral** with only one pair of **parallel** sides

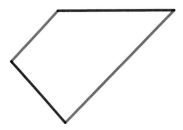

tree diagram: A way to record and count all combinations of events, using lines to form branches (e.g., This tree diagram shows all the things that can happen if you flip a coin twice.)

1st flip   2nd flip

trend: The general direction of data presented in a graph; the data can increase, decrease, or stay the same over time

increase

decrease

stays about the same

triangle: A **polygon** with three sides

turn: See **rotation**. (See also **transformation**.)

unit rate: A comparison of two quantities where the second one is described as 1 unit (e.g., 30 km in 1 h or 4 tomatoes for $1)

unlikely outcome: A result that has little chance of occurring (e.g., If you roll a die with a 3 on all the faces but one, rolling a number other than 3 is an unlikely outcome.)

variable: A quantity that varies or changes; a variable is often represented by a letter or a symbol

Venn diagram: A way of showing the relationship(s) between collections of objects or numbers

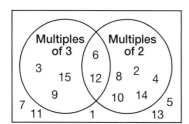

This Venn diagram shows that 6 and 12 are both **multiples** of 2 and multiples of 3.

vertex (plural is vertices): The point at the corner of an angle or a shape (e.g., A cube has eight vertices. An angle has one vertex.)

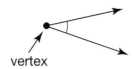

vertex

vertical line: A line that goes up and down a page parallel to the side edge, or straight up and down from the floor

volume: The amount of space occupied by an object; a common unit of measurement is the **cubic centimetre (cm³)**

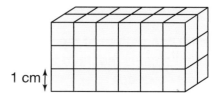

The volume of this box is 36 cm³.

whole numbers: The counting numbers that begin at 0 and continue forever; 0, 1, 2, 3, …

Selected Answers to Problem Bank Questions

Chapter 1, pp. 30–31
2. b) 5.5 min
3. b) $130
4. b) 150; 216
7. 12 586 269 025
8. 620
9. 24

Chapter 2, p. 62
2. about 33 000
5. 11 272
6. 986 742
7. 33

Chapter 3, p. 96
6. a) 8; 7

Chapter 4, pp. 131–132
1. a) 1796
   b) 1867
   c) 1899
4. a) 3990
   b) 406
5. 6
6. a) 9.50 kg
9. 0.555 kg
10. 6.399

Chapter 5, p. 158
1. about 800 g
2. 99 m
5. 13 cm
6. 38 cm
7. 3 m; 4 m; 5.33 m

Chapter 6, p. 201
1. c) 36
3. a) 1783
   b) 2225
4. 3 pairs
5. 24 quarters; 16 half dollars
6. 18 000
9. 25 kg
10. $10 250
13. a) 85 cm
    b) 158 950 cm^2

Chapter 8, p. 258
1. 10 000
2. 800
4. 7560
5. 9

Chapter 9, p. 286
5. 2000

Chapter 10, p. 315
4. b) 3.5 m

Chapter 11, p. 344
2. a) 1240 cm^2
   b) 136 cm^2

Chapter 12, p. 382
4. 8 L of soda
5. b) 80%
6. $16

Chapter 13, p. 409–410
7. $\frac{3}{9}$
10. $\frac{4}{28}$

Index

Adding
  in checking answers to division, 203
  decimal numbers, 120, 135
  decimals, 122–123, 127
  decimals, by renaming, 299
  whole numbers, 108, 112–113
Angle measures
  estimating, 210–211
Angles
  acute, 209
  estimating size of, 216
  obtuse, 209
  right, 209
  sizes in construction of parallelograms, 229
  straight, 209
Area
  estimating, 244
  measurement, and perimeter, 154, 159
  and missing dimensions, 259
  of parallelograms, 259
  of polygons, 252–253, 259
  puzzle, 238–239
  rule for hexagons, 252–253
  rule for parallelograms, 242–244, 248, 252
  rule for rectangles, 248
  rule for triangles, 246–247, 248
Astronomical units, 6
Axes
  horizontal, 72, 74, 78
  vertical, 72, 70

Base, 326
  of parallelograms, 242–243, 246, 259

  of prisms, 326
  of pyramids, 326
  of triangles, 246
  of triangular prisms, 326
Billions, 39
Body relationships, 100–101
Broken-line graphs, 69
Bytes, 44–45

Calculating
  area of polygons, 252–253, 259
  decimal portion of multiple of 10, 276–277
  decimal products, 287
  differences, 108
  the least number, 309
  missing dimensions, 259
  money, 121
  percents, 372–373
  sums, 108
  sums and differences, 108–109, 120–121
  surface are of polyhedrons, 332
  volume of prisms, 326–328
Carroll diagrams, 91–92, 97
Centre of rotation, 418
Chapter tasks
  Chartering Buses, 206
  Creating Nets, 440
  Furnishing a Bedroom, 232
  Growing Up, 290
  Investigating Body Relationships, 100
  Judging the Fairness of a Game, 318
  Mapping out a Biathlon Course, 162
  Painting bids, 348
  Patterns in Your Life, 34
  Placement and Napkin Set, 262

  Reporting numbers, 66
  Running with Terry, 386
  Winning Races, 414
Clockwise (CW), 418
Coins
  flipping, 406
  values, 172–173
Common difference, 9
Communicating
  about triangles, 214–215
  data displays, 88–89
  multi-step problem solving, 128–129
  problem solving, 46–47, 194–195, 278–279
  problem solving in decimal multiplication, 278–279
  using diagrams, 428–429, 438
Comparing
  angle measures, 210–211
  decimals to thousandths, 56–57
  fractions, 356–357, 366
  measurements with different units, 144–145
  numbers to 1 000 000, 42–43
  percents, fractions, and decimals, 370–371
  probabilities, 390–391
  sets of data, 97
  using mean and median, 84–85
Composite numbers, 166–168, 169
Constructing
  parallelograms, 229
  polygons, 218–220
  polyhedrons, 322
Coordinate grids, 72
Coordinate pairs, 72–73, 74, 78, 79, 82
Counterclockwise (CCW), 418

Creating
cube structures, 336–337, 340
designs by transformations, 432–433, 438
division problems, 194–195
line graphs, 74–76
multiplication problems, 194–195
nets, 440
patterns using spreadsheets, 20–21
problem solving, 194–195
scatter plots, 78–80
surveys, 70

Cross-sections, 329

Cubes
polyhedrons from, 334–335

Cube structures
from drawings, 336–337
isometric sketches of, 334, 345
views of, 338–341, 345

Curious math
Alphabet Symmetry, 425
Billions, 39
Changing Parallelograms, 254
Cross-sections, 329
Decimal Equivalents, 281
Dividing Magic Squares, 303
Egyptian Division, 197
Folding Along Diagonals, 221
Lattice Multiplication, 181
Math Magic, 7
Number Reversal, 109
Planes of Symmetry, 341
Random Numbers and Letters, 397
Rice on a Chessboard, 21
Subtracting a Different Way, 117
Telling Stories about Graphs, 77
Triangle Sides, 152

Data
communicating about displays, 88–89
comparing sets of, 97
conclusions from, 88–89

Decagons, 223

Decametres, 142

Decimal hundredths
multiplying by 1000, 268–269

Decimal numbers
adding, 120, 135
subtracting, 120, 135

Decimals
adding, 122–123, 127
adding, by renaming, 299
communicating about problem solving in multiplying, 278–279
comparing, 56–57
dividing, 294–295
dividing by 10, 100, 1000, and 10 000, 306–308, 316
dividing by one-digit numbers, 300–302, 304
dividing by renaming, 58
equivalents, 281
equivalents of fractions, 360, 366
greater of two, 63
missing, 71
modelling, 50–52
multiplying by 10 000, 272
multiplying by single-digit numbers, 272
ordering, 56–57
percents and, 370–371, 383
reading, 50–52
rounding, 54–55
subtracting, 124–126, 127
with three places, 63
writing, 50–52

Decimal tenths
estimating products of, 266–267, 272
multiplying by 1000, 268–269

Decimal thousandths, 50–52, 56–57
multiplying by 1000, 268–269
rounding, 63

Denominators, unlike, 358–359

Designs created by transformations, 416–417, 432–433, 438

Diagonals
folding along, 221
in quadrilaterals, 224–225

Diagrams, communicating with, 428–429, 438

Differences
calculating, 108
estimating, 110–111

Dimensions, 240–241, 250–251
missing, 259

Dividing
decimals, 294–295
decimals by one-digit numbers, 300–302, 304
Egyptian division, 197
by hundreds, 186–187, 203
magic squares, 303
money, 298–299
by 100, 306–308, 316
by 1000, 306–308, 316
by 1000 and 10 000, 184–185, 306–308
problem solving in, 194–195
by 10, 306–308, 316
by tens, 186–187, 203
by 10 000, 306–308
by two-digit numbers, 190–192
using multiplication and addition in checking answers, 203

Double bar graphs, 69

Equal expressions, 24–25
Equals sign, 24, 32
Equations
  equals sign in, 24, 32
  missing terms in, 24–25
  variables in, 26–29
Equilateral triangles, 158, 209, 215, 229, 294–295
Equivalent ratios, 364–366
Eratosthenes, 169
Estimate the Range (game), 297
Estimating
  angle measures, 210–211
  area, 244
  area of polygons, 244
  differences, 110–111
  length of sides of polygons, 294–296
  lengths of sides of squares, 294–295
  percents, 372–373
  products, 176–177, 266–267, 272
  products of decimal tenths, 266–267, 272
  products of money amounts, 266–267
  quotients, 188–189, 203, 294–296, 304
  size of angles, 216
  sums, 110–111
Experimental probability
  theoretical probability and, 404–405, 411
Explicit pattern rules, 9, 18
Expressions
  balanced, 24
  changes in value of, 196–197
  equal, 24–25
  values in, 32
  variables in, 12–13, 18

Factors, 166–168, 182
Farming, 235
Five-digit numbers
  subtracting four-digit numbers from, 118
Four-digit numbers
  dividing by two-digit numbers, 190–192
  multiplying two-digit numbers by, 178–180
  subtracting from five-digit numbers, 118
Fractions
  comparing, 356–357, 366
  decimal equivalents of, 360, 366
  ordering, 356–357, 366
  percents and, 370–371, 383
  and unlike denominators, 358–359
  visualizing, on number line, 391
Frequency, 108

Graphic organizers, 90–92
Graphs, 70
  broken-line, 69
  double bar, 69
  intervals on, 86
  line, 74–76, 77, 82, 87
  representing patterns on, 14–16, 18
  scales on, 87
  telling stories about, 77
Greater than (sign), 42
Guess and test strategy, 376–377

Halving and doubling
  multiplying by, 173

Height
  of parallelograms, 242–243, 248, 259
  of triangles, 246, 248
Hexagons, 225, 295
  area rule for, 252–253
Horizontal axis, 72, 74, 78
Hundreds
  dividing by, 186–187, 203
  multiplying by, 174–175, 182
Hurdles, distance between, 300–302
Hypotheses, 214–215

Inequality sign, 42
Intervals
  on graphs, 86
  organizing trials into, 69
Isometric sketches, 334, 345
Isosceles triangles, 209, 215

Kilobytes, 44–45
Kilometres, and leagues, 87

Leagues, 87
Length
  ancient units of, 292–293
  choice of units of, 142, 148
  measurements with different units of, 144–145, 148
  measuring, 142
  renaming measurements of, 148
  of sides, 216
  of sides of parallelograms, 229
  of sides of polygons, 218–220, 294–296
  of sides of squares, 294–295

of sides of triangles, 216
of time, 155
Less than (sign), 42
Linear metric units
   square and, 240–241
Line graphs, 74–76, 77, 82, 87
Lines, Lines Lines (game), 153
Lines of symmetry, 215, 425
   sorting polygons by,
     222–223, 229
Logical reasoning, 150–151,
   159

Magic sum, 303
Math games
   Calculate the Least
     Number, 309
   Close as You Can, 53
   Coin Products, 193
   Estimate the Range, 297
   4 in a Row, 81
   Lines, Lines Lines, 153
   Mental Math with Money,
     112
   No Tails Please!, 406
   "Odd and Even", 108
   Race to 50, 282
   Ratio Concentration, 378
   Spin and Factor, 166
   Tic-Tac-Toe, 81
   Who Am I?, 17
Mean, 84–85, 97
Measuring
   length, 142
   rotations, 420–421
Median, 84–85, 97
Megabytes, 44–45
Mental imagery
   Drawing Faces of
     Polyhedrons, 323
   Identifying Transformations,
     434
   Visualizing Fractions on a
     Number Line, 391
   Visualizing Symmetrical
     Shapes, 213

Mental math, 11, 58, 71
   in adding, 135
   Adding Decimals by
     Renaming, 299
   for calculating sums and
     differences, 108–109,
     120–121
   in dividing by 1000 and
     10 000, 184–185
   Halving and Doubling to
     Multiply, 173
   with money, 112
   in multiplying, 274–275, 287
   in subtracting, 135
   Using Factors to Multiply,
     361
   Using Whole Numbers to
     Add and Subtract
     Decimals, 127
Metric relationships, 144–145
Mirror symmetry, 341
Modelling
   decimal thousandths, 50–52
   ratios, 362–363
Models, making of, 330–331
Money
   calculating change, 121
   dividing, 298–299
   estimating products of
     amounts, 266–267
Multiples, 182
   identifying, 170–171
Multiplying
   in checking answers to
     division, 203
   choosing methods, 280–281
   creating and explaining
     how to solve problems,
     194–195
   decimals by single-digit
     numbers, 272
   decimals by 10 000, 272
   in estimating quotients,
     188–189
   by 5 and 50, 267
   by halving and doubling, 173
   by hundreds, 174–175, 182
   lattice, 181

multiples of ten by tenths,
   276–277
by 1000, 268–269
by pairing, 11
symbol used in
   spreadsheets, 20
by tens, 182
by 10 000, 268–269
tenths by whole numbers,
   270–271
by two-digit numbers,
   178–180
using factors, 361
by 0.1, 274–275, 287
by 0.01, 274–275, 287
by 0.001, 274–275, 287
Multi-step problems
   communicating about
     solving, 128–129

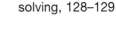

Nets
   creating, 440
   solving puzzles, 320–321
Number clues, 36–37
Numbers
   greater than 100 000, 40–41
   to 1 000 000, 38–39, 42–43
   reading, 40–41
   renaming, 44–45, 48
   six-digit, 48
   term, 4–5
   writing, 40–41

Obtuse angle, 209
Obtuse triangle, 209, 247
One-digit numbers
   dividing decimals by,
     300–302, 304
Operations, order of, 196–197
Opposite sides, 213
Ordering
   decimals, 56–57
   decimals to thousandths,
     56–57

fractions, 356–357, 366
numbers to 1 000 000, 42–43
percents, fractions, and decimals, 370–371
Order of rotational symmetry, 423, 425
Origin, 72
Outcomes, 400

Pairing, to multiply, 11
Parallelograms, 218–220, 224, 229
area of, 259
area rule for, 242–244, 248, 252
base of, 246, 259
changing, 254
height of, 248, 259
triangles and, 245
Pattern rules, 4–6
explicit, 9, 18
recursive, 9, 18
relationship, 8–11
spreadsheets and, 32
term values and, 18
Patterns, 34
graphing of, 18
on graphs, 14–16, 18
of growth, 2–3
represented in tables, 14
represented on graphs, 14–16
spreadsheets and, 20–21, 32
Pentagons, 220, 229, 295
Percents, 368, 383
calculating, 372–373
and decimals, 370–371, 383
in describing probabilities, 392–393, 398
estimating, 372–373
and fractions, 370–371, 383
as special ratios, 368–369

Perimeters
and area measurements, 154, 159
of polygons, 146–147, 148, 154, 159
Periods, 40
Perpendicular lines, 242–243
Pictures, memorization of, 68–69
Pixels, 174
Planes of symmetry, 341
Plots, scatter, 78–80, 82, 88
Plotting, 72, 82
Polygons
area of, 259
areas of, 252–253
congruent, 322
constructing, 218–220
estimating area of, 244
estimating length of sides, 294–296
"mystery", 208–209
perimeters of, 146–147, 148, 154, 159
snowflake, 158
sorting by line symmetry, 222–223, 229
Polyhedrons, 322
constructing, 322
from cubes, 334–335
drawing faces of, 323
surface area of, 324–325, 332
visualizing, 322
Predictions, 78, 79
Prime numbers, 166–168, 169
Prisms, 322. See also Rectangular prisms; Triangular prisms
base of, 326
volume of, 332
Probabilities
experiments, 390–391, 398
and random numbers and letters, 397
using percents to describe, 392–393, 398

Probability lines, 391
Problem solving
communicating about, 46–47, 194–195, 278–279
communicating about multi-step, 128–129
by conducting experiments, 394–396
creating, 194–195
explaining thinking in, 46–47
guess and test strategy, 376–377
by making models, 330–331
in open sentences, 250–251
using logical reasoning, 150–151, 159
by using simpler problems, 22–23
by working backward, 310–311
Products, estimating of, 176–177, 266–267, 272
Pyramids, 322
base of, 326

Quadrilaterals, 219, 220, 221
properties of, 224–225
Quotients
estimating, 188–189, 203, 294–296, 304

Random, 394
choosing numbers and letters, 397
Ratios, 362–363
concentrations, 378
equivalent, 365–365, 366
percents as, 368–369

Rectangles
  area rule for, 248
  diagonals in, 224
Rectangular prisms
  surface area of, 324–325
  volume of, 326–328
Recursive pattern rules, 9, 18
Renaming
  in adding decimals, 299
  and dividing by tens and hundreds, 186–187
  dividing decimals by, 58
  length measurements, 148
  in multiplying by 5 and 50, 267
  numbers, 44–45
Representing
  patterns in tables, 14
  patterns on graphs, 14–16, 18
  variables by letters, 13
Rhombuses, 220, 224
Right angles, 209
Right triangles, 209, 247
Rotational symmetry, 422–424, 425, 426
Rotations
  describing, 418–419, 426
  measuring, 420–421
Rounding, in estimating quotients, 188–189

Scale diagrams, 218–220
Scalene triangles, 209, 215
Scatter plots, 78–80, 82, 88
Side lengths, 216
  in constructing parallelograms, 229
  in constructing polygons, 218–220
Sides, 152
  number of, and lines of symmetry of polygons, 229
  opposite, 213
Single-digit numbers
  multiplying decimals by, 272

Six-digit numbers, 48
Skill-testing questions with more than one answer, 196–197
Sorting
  polygons by lines of symmetry, 222–223, 229
  quadrilaterals by properties, 224–225
Spreadsheets, 70
  and pattern rules, 32
  patterns and, 20–21
Square metres vs. square centimetres, 248
Square metric units
  linear and, 240–241
Squares, 225
  diagonals in, 224
  estimating lengths of sides, 294–295
Straight angles, 209
Subtracting, 117
  decimal numbers, 120, 135
  decimals, 124–126, 127
  four-digit from five-digit numbers, 118
  in mental math, 135
  whole numbers, 108, 114–116
Sums
  calculating, 108
  estimating, 110–111
Surface area, 324
  of polyhedrons, 324–325, 332
  of rectangular prisms, 324–325
  of triangular prisms, 324–325
Surveys, 70
Symmetrical shapes,
  visualizing, 213
Symmetry
  alphabet, 425
  line, 222–223, 229
  lines of, 215, 425
  mirror, 341
  planes of, 341
  rotational, 422–424, 425, 426

Tables, 14
Tally charts, 70
Tangram puzzles, 245
Tenths
  multiplying by whole numbers, 270–271
  multiplying multiples of ten by, 276–277
Term numbers, 4–5
Term values, pattern rules and, 18
Theoretical probability, 400–401
  experimental probability and, 404–405, 411
  tree diagrams for, 402–403, 411
Three-digit numbers
  adding, 112–113, 118
Time, lengths of, 155
Tonnes, 176
Transformations, 426
  creating designs by using, 416–417, 432–433, 438
  exploring with technology, 430–431
  identifying, 434
  and number patterns, 430–431
Trapezoids, 224
Tree diagrams, 402–403, 411
Triangles, 209
  acute, 209, 247
  angle measures and side lengths, 216
  angle relationships of, 212–213
  area rule for, 246–247, 248
  base of, 246
  communicating about, 214–215
  congruent, 221, 245
  equilateral, 158, 209, 215, 229, 294–295
  height of, 246, 248
  isosceles, 209, 215

obtuse, 209, 247
opposite sides of, 213
and parallelograms, 245
properties of, 212–213
right, 209, 247
scalene, 209, 215
sides of, 152, 212–213
Triangular prisms
base of, 326
surface area of, 324–325
volume of, 326–328
Triathlons, 162
Trillions, 39
Troy ounces, 138
Two-digit numbers
dividing by, 190–192
dividing four-digit numbers by, 190–192
multiplying by, 178–180

Unit rates, 374–375, 383
Units of length
ancient, 292–293
choice of, 142, 148
measurements using different units, 144–145, 148

Variables
in equations, 26–29
in expressions, 12–13, 18
represented by letters, 13
Venn diagrams, 90–91, 223

Vertical axis, 72, 78
Visualizing
fractions on number line, 391
polyhedrons, 322
symmetrical shapes, 213
Volume
of prisms, 332
of rectangular prisms, 326–328
of triangular prisms, 326–328
Writing
decimal thousandths, 50–52
numbers, 40–41

Credits

This page constitutes an extension of the copyright page. We have made every effort to trace the ownership of all copyrighted material and to secure permission from copyright holders. In the event of any question arising as to the use of any material, we will be pleased to make the necessary corrections in future printings. Thanks are due to the following authors, publishers, and agents for permission to use the material indicated.

Front and Back cover: T. Kitchin/First Light

Table of Contents, Page vii (top left): EPA/GERO BRELOER/Landov; Page viii: Robert Laberge/Getty Images; Page xi (top): © Kennan Ward/Corbis; (bottom left) Fred Chartrand/CP Picture Archive; Page x (top left): CP Photo Archive; Page xi (middle right): Andrew Vaughn/CP Picture Archive; Page xii (top right): Photodisc Green/Getty Images; (bottom left) Courtesy of Ashrita Furman; Page xv (bottom right) Michael Mahaovlich/Masterfile; Page xvi (top right): T. Kitchin/First Light, (bottom left) Courtesy of Brian Atkinson: Page xviii (left): © Bettman/Corbis

Chapter 1, Page 4 main: © Corel, inset: NASA; Page 5: NASA; Page 6: Jupiter Images; Page 19: NASA; Page 30: Jupiter Images; Page 31: EPA/GERO BRELOER/Landov; Page 33: © Picture Arts/Corbis

Chapter 2 Opener, Page 35 left to right: Jupiter Images, NASA, © Comstock/Alamy, Shaun Best/Reuters/Landov, © Corbis, Page 40 left: © Roger Ressmeyer/Corbis, right: © Roger Ressmeyer/Corbis; Page 41: © Roger Ressmeyer/Corbis; Page 45: Jupiter Images; Page 46: © ER Productions/Corbis; Page 52: World Perspectives/Photographer's Choice/Getty Images; Page 56: Robert Laberge/Getty Images; Page 57: Scott Simms/Foodpix/Getty Images

Chapter 3 Page 68 (1st 4 images): Photodisc/Getty, Stockbyte/PictureQuest, Ryan McVay/Photodisc Green/Getty Images, Ryan McVay/Photodisc Green/Getty Images, Jupiter Images, © Corbis Royalty Free, Alask Stock LLC/Alamy, © Corbis Royalty Free, Jupiter Images, Burke/Triolo/Brand X Pictures/Getty Images; S.Solum/Photolink/Photodisc Green/Getty Images; Photodisc/Getty Images, Photodisc/Getty Images, Brand X Pictures/Getty Images, Photodisc/Getty Images, Photodisc/Getty Images, Photodisc/Getty Images; Page 78: Fred Chartrand/CP Picture Archive; Page 84: © Swift/Vanuga Images/Corbis, © Eric & David Hoskings/Corbis, T. Kitchin/First Light; Page 85: Alan & Sandy Carey/Photodisc Green/Getty Images, Digital Vision/Getty Images; © Royalty-free/Corbis; Page 96: © Kennan Ward/Corbis; Page 103: © Natalie Fobes/Corbis, © David A. Northcott/Corbis, © David A. Northcott/Corbis, © Rob C. Nunnington; Gallo Images/Corbis; Page 104: Altrendo/Getty Images, © Buddy Mays/Corbis;

Chapter 4 page 105: Nikos Paraschos/EPA/Landov, Brett Coomer/Associated Press, Ryan Remiorz/CP Picture Archive, Jean-Baptiste Benquent/CP Picture Archive; Page 106: © Tony Freeman/Photo Edit; Page 110: UPI Photo/Landov; Page 111: NOAA; Page 116: Chryssa Panousiadou/EPA/Landov; Page 126: Heine Ruckeman/UPI/ Landov; Page 133: © Bettman/Corbis; Page 138: CP Photo Archive

Chapter 5 Page 142 top: Gunter Marx/Alamy, middle: middle: © Tony Wharton; Frank Lane Picture Agency/Corbis, bottom: © Corel; Page 143: John A. Rizzo/Photodisc Green/Getty Images; Page 148: top left: © Image Farm Inc./Alamy, top middle: Photodisc/Getty Images, top right: Jupiter images, lower right: © Superstock/Alamy; Page 149: Andrew Ward/Life File/Photodisc Green/Getty Images; Page 156: Andrew Vaughn/CP Picture Archive

Chapter 6 Opener, Page 163 main: Brand X Pictures/Alamy, inset: Michael Melford/The Image Bank/Getty Images; Page 165: Geoff Howe/CP Picture Archive; Page 170: Jason T. Ware/Photo Researchers; Page 171: Chris Cook/Photo Researchers; Page 172: Geoff Howe/CP Picture Archive ; Page 176: © Lester Lefkowitz/Corbis; Page 178: J.P. Moczulski/CP Picture Archive; Page 180: Courtesy of Ashrita Furman; Page 181: S. Solum/Photolink/Photodisc Blue/Getty Images; Page 190: © W. Cody/Corbis; Page 199: © Creatas/PictureQuest; Page 200: Guy Crittenden/Index Stock; Page 201 top: © John T. Fowler/Alamy, middle: Photodisc Green/Getty Images, bottom: © Corbis Royalty Free; Page 204: © Royalty Free/Corbis; Page 205: © Carl & Ann Purcell/Corbis

Chapter 7 Page 233 top: © Douglas Peebles/Corbis, bottom: Rob Nunnington/Oxford Scientific Films; Page 235 top: Michael Lea/CP Picture Archive, middle: Bernd Fuchs/First Light; Page 236: top: 4-H Canada, bottom: Photography by Catherine McKeough/4-H Canada

Chapter 8 Opener, Page 237: © Image State/Alamy; Page 247: © Kelly-Mooney Photography/Corbis

Chapter 9 Opener, Page 263: Science North — Moriyama and Teshima Architects; Page 268: Michael Mahaovlich/Masterfile; Page 269: Jupiter Images; Page 283: Reuters/Richard Chung/Landov; Page 290: top: Jupiter Images, bottom: © Corel

Chapter 10 Opener, Page 291: © Martin Cushen/Alamy; Page 305: © T. Kitchin/First Light; Page 306: top left: Jupiter Images, top middle: Courtesy of Seevirtual360.com. Photo originated from the Village of Alert Bay, BC, top right: Courtesy of Brian Atkinson, lower right: © Ed Darak/Darack.com/Alamy; Page 308 left: Don Denton/CP Picture Archive, middle: © Gunter Marx Photography/Corbis, right: © britishcolumbiaphotos.com/Alamy

Chapter 11 Page 341: Koichi Mamoshicka/Getty Images; Page 351 top: Jupiter Images, middle: Ron Watts/First Light, bottom: Public Works and Government Services; Page 352 top and bottom: © Corel

Chapter 12 Page 364: © Guy Motil/Corbis; Page 370: © Benjamin Lowy/Corbis; Page 386: © Bettman/Corbis

Chapter 14 Opener, Page 415: © fStop/Alamy; Page 418: BGA Publishing USA © Beau Gardner